American Institute of Chemical Engineers

Topical Conference on Using Process Integration Technology in Industry to Address Environmental Issues 2004

Held at the 2004 AIChE Spring National Meeting

April 25-29, 2004
New Orleans, Louisiana, USA

Printed from e-media with permission by:

Curran Associates, Inc.
57 Morehouse Lane
Red Hook, NY 12571
www.proceedings.com

ISBN: 978-1-60423-480-0

American Institute of Chemical Engineers

Topical Conference on Using Process Integration Technology in Industry
to Address Environmental Issues
2004

TABLE OF CONTENTS

Design and Development of Eco-Industrial Parks

Authors

Dennis Spriggs, Matrix Process Integration
Ernie Lowe, Indigo Development, Sustainable Systems Inc.
Jill Watz, Lawrence Livermore National Laboratory
Mahmoud El-Halwagi, Texas A&M University, Department of Chemical Engineering

Keywords: Sustainable, design, integration, eco-industrial parks

Prepared for Presentation at the AIChE Spring Meeting, New Orleans, April 2004,

Paper # 109a

Unpublished

Abstract

It is well accepted that resource conservation, pollution prevention and cost reduction are all compatible objectives and have received considerable attention at the local (e.g. plant) level. The result has been increased efficiencies, greater productivity and enhanced environmental performance. In spite of this, global economies continue to deplete finite resources at an increasing and alarming rate. In this paper we discuss the challenges of a more sustainable economy in which the demands for products and services are met while simultaneously decreasing the demand for finite resources. We describe how an integrated approach at a macro (multi-plant/business) level can work effectively to generate efficiencies that produce step-change improvements in sustainability. We describe the role of eco-industrial parks and eco-industrial networks and discuss their application to the Dalian Development Zone in China.

Introduction

The Circular or Recycling Economy

There is a growing recognition that industrialized nations are beginning to "hit the wall" with respect to once-through utilization of non-renewable resources. This has led to the concept of the circular economy in which such resources are recycled/reused to minimize their depletion. In addition, renewable resources are consumed at a rate compatible with their rate of renewal. These actions in combination contribute to long term sustainable development in terms of resource use.

Approaches

The broad concept of the circular or recycling economy has been discussed, developed and put into action in one form or another in many locations and under many different circumstances. These initiatives have been both formal/planned and informal/self organizing.

At one extreme are activities that might be characterized as "business as usual". These informal/self-organizing actions have grown naturally out of business taking its natural course with integration and resource conservation following as a consequence of cost reduction or profit improvement. For example:

- Cogeneration is well recognized as a way to promote energy efficiency and achieve cost reduction by combining power generation with process heating. One company owns the cogeneration facility while another serves as the "host" for a portion of the output.
- Along the US Gulf Coast, there are numerous examples of one company providing a C2+ rich fuel gas stream to another where valuable

components are removed and lean fuel gas returned to the original company (Lyondell Citgo, Exxon Mobil, Shell and others). In Corpus Christi, Texas, this is done on a broad scale by several refineries with a 3rd party providing the processing equipment to remove and market the valuable components in the fuel gas.

- Within individual companies, businesses are co-hosted at one site because of mass or energy synergies – one process might be endothermic and another exothermic.

At the other extreme, the Chinese Circular Economy Initiative integrates cleaner production and industrial ecology in a planned fashion. This broad approach encompasses industrial firms, networks or chains of firms, eco-industrial parks and regional infrastructure. The goal is to support efficient resource utilization and reduce toxic emissions. There are three basic levels of action:

- At the individual firm level, managers are directed to seek much higher efficiency through the three Rs of cleaner production: *reduce* consumption of resources and emission of pollutants/waste, *reuse* resources, and *recycle* by-products. Specific targets are set for intensive resource users and polluters
- The second level is to reuse and recycle resources within industrial parks and clustered or chained industries, so that resources will circulate fully in the local production system. This is equivalent to the US tactic of by-product exchange. Eco-industrial parks are specifically called for in these plans, and are envisioned to be environmentally sensitive industrial development projects with a particular mission of hosting resource recovery firms and others that support the circular economy. (The US Environmental Protection Agency supported the first major conceptualization of the EIP in 1994-5.)
- The third level is to integrate different production and consumption systems in a region so the resources circulate among industries and urban systems. This level requires development of municipal or regional by-product collection, storage, processing, and distribution systems.

The Eco-Industrial Park

It is clear that substantial progress toward sustainable development will occur only if we break out of the "business as usual mode". Toward that end, we distinguish three basic categories of eco-industrial projects:

- Eco-industrial Park or Estate (EIP): This can be defined as a cluster of processing and other facilities together with supporting infrastructure which create the circular economy and promote sustainable development. Such a park will be developed and managed as a real estate development, seeking high environmental, economic, and social benefits as well as business excellence. It will implement conditions for participation and will recruit members who in total achieve the desired objective. This park will be bottom line driven and the tenants selected in such a way that economic performance increases as the level of integration and, therefore, efficiency increases. Participation will provide clear economic and other advantages over the current stand-alone processing model.

- By-Product Exchange (BPX): This is an arrangement whereby a set of companies seeks to utilize each other's by-products (energy, water, and materials) rather than disposing of them as waste. One person's waste is another's raw material.
- Eco-Industrial Network (EIN): This is an extension of the EIP, comprised of a set of companies collaborating to improve their environmental, social, and economic performance in a region. This can include a network of stand alone companies and/or EIPs.

The Challenges

There are two classes of challenges in designing and developing integration on this macro level – the first is technical/economic and the other organizational/commercial/political.

- The Technical/Economic Challenge: The main issue is how to integrate mass and energy flows economically - both locally within a processing unit and globally among many processing units and even companies. This goes significantly beyond the alliances that form naturally between/among companies. The challenge within an eco-industrial park is to promote such integration on many fronts, infrastructure being only one. Products, byproducts and even intermediate material streams must also enter into the overall integration picture. And, the result must be highly attractive economically as well as achieving mass and energy balance.

- The organizational/commercial/political challenge: In many ways these present the largest challenge. It typically is not in anyone's job description to consider integration on this level. Further, such integration must occur across multiple business entities/companies, greatly complicating the organizational job. There are knotty issues that must be resolved: What level of integration should be promoted among individual companies? Who owns what production units? Who owns the infrastructure? What are the commercial arrangements among the integrated parties? What are the regulatory and legal obstacles to this type of integration and what types of changes are required to provide incentives for greater integration? Are tax or other incentives required? The challenge becomes even more difficult as business conditions change over time, requiring the cluster of companies to be flexible and agile and respond quickly to changing market demands. Leaving this to self organization, on the one hand, is likely to fail and on the other, companies typically don't want regulators or economic development agencies to meddle in their commercial affairs.

Technical and Economic Considerations

One key to solving this integration puzzle is a sound framework for integrated design. For this we can call on the well developed and proven energy and mass integration methods and tools that have been developed for single unit process design. We extend these methods to broader networks of processes and businesses (e.g. EIPs) by drawing the circle larger. These methods must:

- Be based on fundamental principles
- Employ mathematical, thermodynamic and hierarchical methods
- Provide a framework for inter-plant mass and energy integration
- Use systematic techniques to extract optimum solutions
- Be flexible enough to allow evolution over time

The foregoing requirements can be uniquely matched by extending newly evolved process integration techniques (Dunn and El-Halwagi, 2003; El-Halwagi and Spriggs, 1998; El-Halwagi, 1997).

Process integration is a holistic approach to the design and operation of complex systems. It can be broadly classified into mass integration and energy integration. Mass integration provides a systematic way of tracking species throughout the system and developing optimum strategies for allocation, separation, and generation of species and streams. It also provides a way of targeting the performance of the system. The first step in design is to develop a structural representation of the EIP. This is shown in Figure 1. Each plant receives fresh feeds, material utilities, and yields products, byproducts, and spent materials (the arrows in and out). Furthermore, each plant may consume and/or produce various forms of energy. A centralized system is constructed to facilitate material and energy processing and exchange. Interception technologies may be used to adjust compositions and properties of the streams and condition them to be acceptable for reuse by other plants. Further, the streams entering the integration problem can be intermediate streams within a plant as well as the terminal streams.

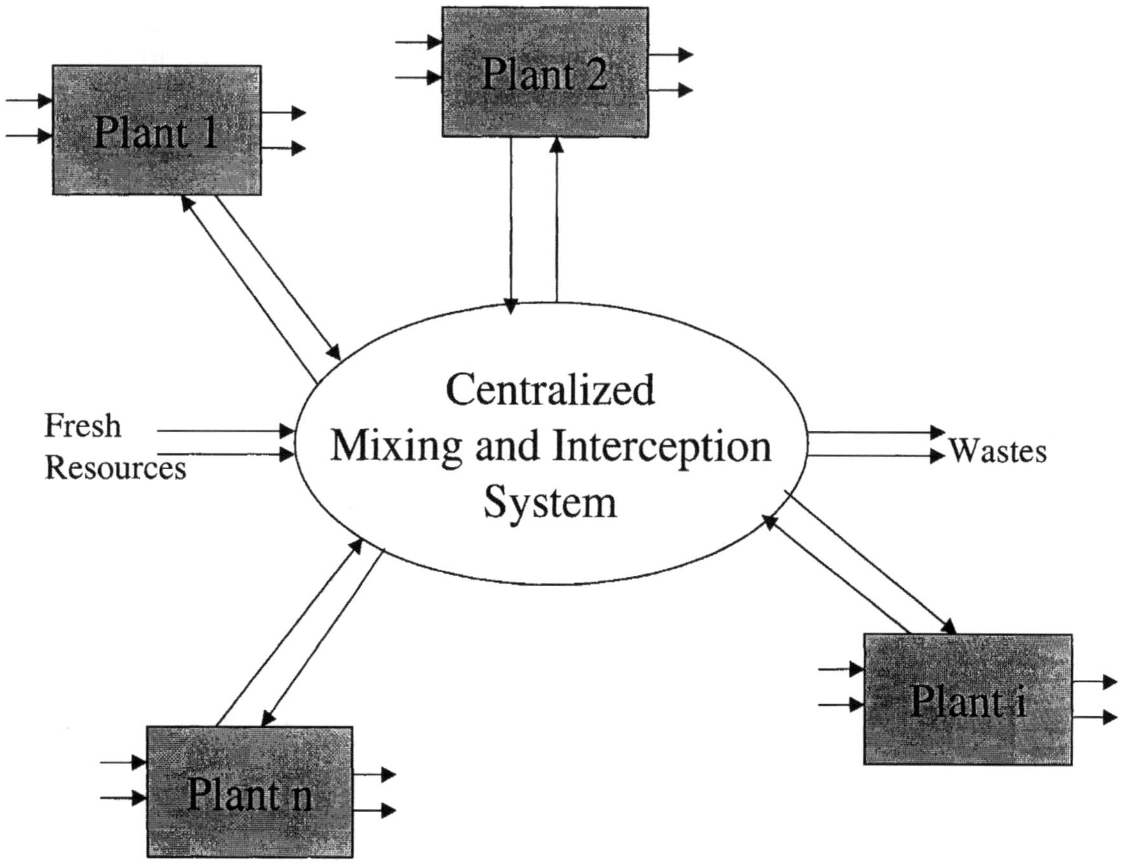

Figure 1
Schematic Representation of Plants and Centralized Mixing/Interception Network

Using this representation, one can develop systematic techniques based on process integration principles. For instance, one strategy for EIP's is to allow byproduct and material exchange (Lowe, 1997, 2000). The problem may be stated as follows:

Given a number of processing facilities each with a certain demand for material consumption and certain production of byproducts and discharges, develop a material exchange network that minimizes the usage of fresh resources and terminal waste discharge.

To solve this problem, data must first be collected on the process requirements and discharges (e.g., flow rate, composition, etc.). Using the material reuse pinch analysis (El-Halwagi et al., 2003), we can determine rigorous targets for minimizing the usage of fresh resources and for minimizing the discharge of terminal wastes. The pinch analysis also determines the optimal strategies for allocating the byproduct and waste streams to process units for effective recycle, reuse, or utilization. The approach can also be used to suggest process design changes in each plant so as to improve the overall performance of the material exchange network. Figure 2 shows the material reuse pinch representation.

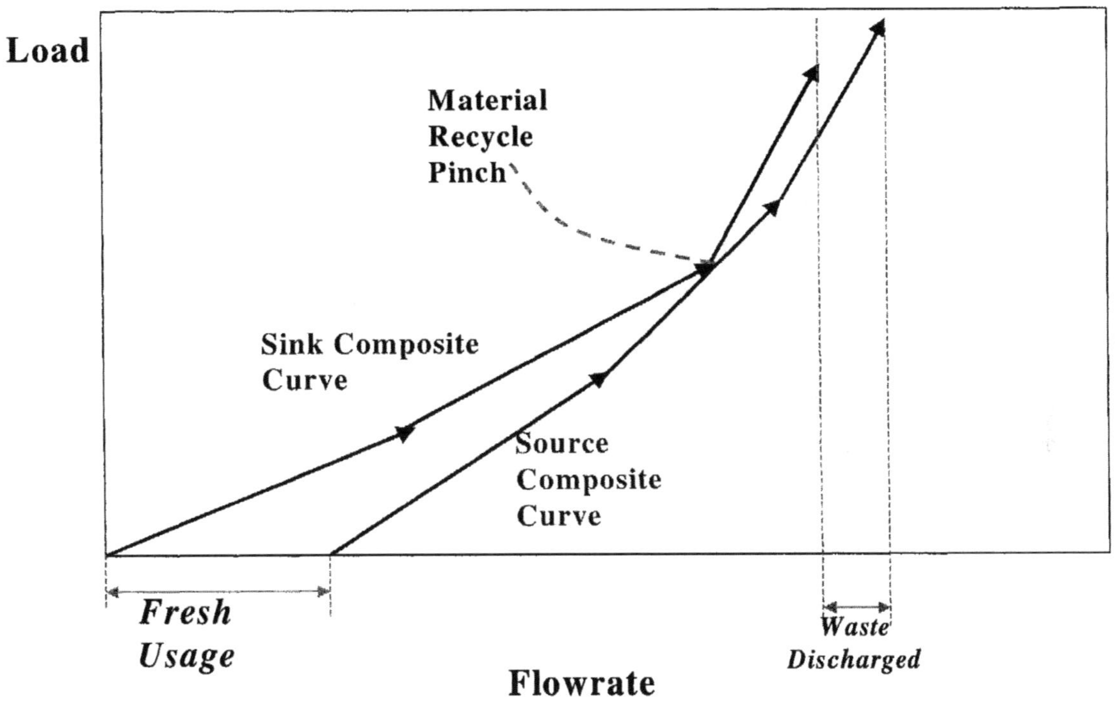

Figure 2

Identifying Targets for Material Exchange Networks Using Pinch Analysis (El-Halwagi et al., 2003)

Additional Technical challenges

In addition to the issues just addressed, there are additional technical challenges that should be the focus of future research. These include:

- Uncertainty: In designing EIPs care must be given to the uncertainty in stream data, forecast of plant performance, and even whether a business will continue in operation.
- Buildup with Closure: As systems move into more integrated modes, material buildup can become a problem and must be accounted for.

- Simultaneous Heat and Mass Exchange: In some cases, it is difficult to treat the issues of mass and heat exchange separately. Simultaneous approaches are needed.
- Dynamic Fluctuations: EIPs must be designed to handle internal and external disturbances all the time.
- Interface with Social Aspects: Techno-economic aspects must be coordinated with social factors. This poses a difficulty in modeling because of the challenges in quantifying and reconciling several human and social issues. One approach is to use environmental biocomplexity as an overarching framework (El-Halwagi, 2003).
- Reliability: Because of the high nature of interconnectedness, the design of an EIP requires the incorporation of backup contingencies in case flows or operations are disrupted in part of the system and fail to contribute their designated function (Cote and Cohen-Rosenthal, 1998).
- Integration-Conducive Landscape: The EIP layout, landscape architecture, and infrastructure support must be designed in a way that is consistent with the overall integration objectives (e.g., Todd et al., 2003; Carr et al., 1998; Grant, 1997)
- Incorporating Geographic Information: Relevant geographic information systems should be utilized to guide the integration and future growth of the EIP (Ozyurt and Reallf, 2001)

Application to the Dalian Development Zone

The Dalian Development Zone (DDZ) is north of the Chinese City of Dalian and located on the coast of Liaoning Province, opposite North Korea. The Zone is essentially a new town of 430 square kilometers, with approximately 1,500 factories, residential areas and a commercial district with hotels and high rise office buildings. Industries include petrochemical, electronics, computers, food processing, and auto parts, among many others. To date, approximately 30 square kilometers have been developed.

DDZ authorities have taken initial steps toward the development of a Dalian Eco-Industrial Park as required by the Liaoning Circular Economy Plan (LCEP). This EIP will seek tenants whose products and services will support the DDZ, its companies, and other firms in meeting the requirements of the Plan. Sustainable Systems Inc. is negotiating an agreement to support planning for the site including buildings, infrastructure and management systems. It will also recruit tenants to the EIP and to land elsewhere in the Zone. The DDZ Planning Bureau has reserved 83 hectares of land situated at the gateway to a newly expanded area of the Zone to house the EIP.

Conceptual Framework

Under the sponsorship of Sustainable Systems, the authors have begun the initial conceptual and design work for the EIP. The conceptual framework upon which the design will be based includes the following fundamental principles:

- Infrastructure and buildings will be designed, constructed, and managed in accordance with environmentally sound practices, equipment, systems, and materials.
- EIP tenants will agree to follow international standards regarding the minimization of pollution and the reuse, recycling, conversion and disposal of waste within their own facilities.
- Enterprises will use a life cycle framework to optimize their use of materials and to minimize environmental damage over the life of the facilities. Their practices will include recycling and reuse as well as minimizing waste products that can not be used by other enterprises located in the vicinity.
- Synergistic linkage of enterprises will be planned so that the use of materials, energy, water, and transportation are minimized for a given output of production. Where possible, enterprises will exchange by-products not recycled internally.
- Learning within and among tenants is considered to be an essential mission of the EIP. To that end, the EIP will operate as a learning center for the region.
- The EIP will serve as a node for the region's eco-industrial network of operations pursuing sustainable development, responding to the LCEP.
- EIP Management will provide services and leadership to assure that all of these requirements are maintained over the life of the park.

Recruitment

Recruitment will emphasize synergies among the companies and businesses located within the park and with other companies, facilities and urban areas in the DDZ and general area. Emphasis will be placed on recruiting companies in the following categories:

- Core businesses and manufacturing firms: These companies provide products and services for local sale or export. Ideally they will have raw material and energy/utility requirements which are highly synergistic, providing the opportunity for by-product exchange and sharing of common infrastructure.
- Infrastructure firms: These companies will identify and implement equipment and services which provide essential energy and other utilities to the Park tenants. These will include co-generation and district heating and cooling where appropriate, building on the synergistic requirements of the tenants.
- Resource recovery firms: These enterprises produce value from major product and material waste flows. They include niche collection, recycling, equipment re-manufacturing, product and material reuse, bio-energy

generation, composting, etc. These companies will accommodate by-product utilization among other companies within the EIP as well as in the larger zone. (With organic materials constituting 75% of the Dalian municipal waste stream, biomass processing for energy and materials should be one major focus.)

- Alternative energy firms: These tenants would offer products and services to increase efficiency of energy use and would directly serve achieving the objectives of the Liaoning Circular Economy Plan. Renewable energy firms would produce products such as bio-gas and bio-fuel, equipment for wind and solar energy, hydrogen fuel cell transportation and power generation, and energy system integrators. Their presence in the EIP and the DDZ could further a long-range LCEP goal of increasing the efficiency and reducing the environmental burdens of energy generation. R & D and business development in this field could create a platform for emerging sustainable technologies and build the technological know-how required to be competitive.
- Water technology and service companies: These companies would support development of more efficient and integrated water management systems in public infrastructure, commercial buildings, and industrial plants. The regional water crisis makes this a critically important cluster.
- Information technology firms supporting environmental management: These companies would provide products and services for environmental monitoring, modeling and simulation, design, real time feedback systems, and community involvement.
- Companies in the supply chain of major industries: Most of the above companies fit particular niches in the supply chains of international and Chinese private firms and state-owned enterprises. Others, such as comprehensive resource services companies or suppliers of environmentally superior materials or components may also be recruited.

Summary

The Dalian Eco-Industrial Park will have a clear mission to support the business and social community of the DDZ and of the City of Dalian. This support could include marketing packages of products and services to State Owned Enterprises, multinationals, and government operations; facilitating creation of environmental and energy companies outside the EIP; managing "chain connections" or by-product exchanges among industries; and furthering social programs that benefit the citizens of the region.

References

Carr, A. J. P., "Choctaw Eco-Industrial Park: An Ecological Approach to Industrial Land Use Planning and Design", Landscape and Urban Planning, 42, 239-257 (1998)

Cote, R. P. and E. Cohen-Rosenthal, "Designing Eco-Industrial Parks: A Synthesis of Some Experience", J. Cleaner Production, 6, 181-188 (1998)

Dunn, R. F. and M. M. El-Halwagi, "Process Integration Technology Review: Background and Applications in the Chemical Process Industry", J. Chem. Tech. And Biotech, 78(1011-1021), (2003)

El-Halwagi, M. M., "1997, Pollution Prevention through Process Integration: Systematic Design Tools", Academic Press, San Diego.

El-Halwagi, M. M., F. Gabriel, and D. Harell, "Rigorous Graphical Targeting for Resource Conservation via Material Recycle/Reuse Networks", Ind. Eng. Chem. Res., 42, 4319-4328 (2003)

El-Halwagi, M. M., 2003, "Industry and environmental biocomplexity: impact, challenges, and opportunities for multidisciplinary research" J. Clean Technologies and Environmental Policy, 4(3), 135

El-Halwagi, M. M. and H. D. Spriggs, 1998, "Solve design Puzzles with Mass Integration", Chem. Eng. Prog., vol. 94, August, pp. 25-44

Grant, J., "Planning and Designing Industrial Landscapes for Eco-Efficiency", J. Cleaner Prod., 5(1-2), 75-78 (1997)

Lambert, A. J. D. and F. A. Boons, "Eco-Industrial Parks: Stimulating Sustainable Development in Mixed Industrial Parks", Technovation, 22: 471-484 (2002)

Lowe, E. A., "Eco-Industrial Park Handbook for Asian Developing Countries, prepared under a grant from Asian Development Bank, Environmental Department. Sustainable Systems Inc. Oakland, www.indigodev.com/Papers.html (2000)

Lowe, E. A., "Creating Byproduct Resource Exchange: Strategies for Eco-Industrial Parks", J. Cleaner Prod., 5(1-2), 57-65 (1997)

Ozyurt, D. B. and M. J. Realff, "Combining Geographic Information Systems, Process, and Environmental Modeling for Macro Level Pollution Prevention", in Process Design Tools for the Environment, Eds. S. K. Sikdar and M. El-Halwagi, pp. 341-368 (2001), Taylor and Francis, NY.

Todd, J., E. J.G. Brown, and E. Wells, "Ecological Design Applied", Ecological Eng., 20, 421-440 (2003)

Behavior of Trickle Bed Air Biofilter for VOCs Removal: Effect of Non-Use Periods

Daekeun Kim, Zhangli Cai, George A Sorial

Department of Civil and Environmental Engineering, University of Cincinnati, Cincinnati, OH. 45221

Abstract

Biofiltration has recently emerged as an attractive option for controlling volatile organic compounds (VOCs) emission. The objective of this work is to study the effect of non-use periods on biofilter performance. Two different non-use periods were investigated in order to mimic periods of repair of equipment or shutdown during weekend and holidays. In this research, trickle bed air biofilter (TBABs) with synthetic, inorganic media for microbial attachment was employed for VOC control from an air stream. VOC loading varied from 0.70 to 7.03 kg COD/m^3·day.

Consistent long term efficient performance of biofilter strongly depended on biomas control. A periodic in-situ upflow washing (backwashing) was approached in this study. For 0.70, 1.41, and 3.52 kg COD/m^3·day, the biofilter provided the 99 % VOC removal efficiency level by conducting backwashing at a rate of 1 hour per week. When the non-use period strategies were conducted, the overall VOC removal efficiency remained at the 99 % range without backwashing for loading rates of 0.70 and 1.41 kg COD/m^3·day. While for a loading rate of 3.52 kg COD/m^3·day, it dropped occasionally below 90 % due to the accumulation of excess biomass.

The non-use period can be considered as another means of biomass control for low VOC loadings (0.70 and 1.41 kg COD/m^3·day), however, at high VOC loading (>3.52 kg COD/m^3·day), backwashing deemed to be necessary. The biofilter response after restart-up following the non-use period was different as compared to the response after backwashing. The biofilter response after restart-up following the non-use period is a strong function of the active biomass in the system. In this paper, kinetics analysis is presented for comparing the reaction rates for each experimental strategy considered.

1. Introduction

Biofiltration is rapidly becoming a promising air pollution control technology for removing volatile organic compounds (VOCs) present in waste air stream, such as gaseous emissions from chemical process. In a biofiltration process, waste gases containing biodegradable VOCs or inorganic air toxics are degraded to environmentally benign end products by passage through a biologically active medium. In response to the growing concern over VOCs and due to its cost effectiveness, biofilters have been the subject of extensive research since the 1980s.

The concept of the trickle bed air biofilter (TBAB) allows consistent nutrient and pH control for optimizing the waste utilizing kinetics for micro-organisms. It consists of a reactor containing microbes supported on the synthetic medium, with nutrients typically delivered at a minimal liquid flow rate through a nozzle system on top of the bed. The main disadvantage of the TBAB is clogging due to excess biomass formation in the biofilter system. Periodic in-situ upflow washing, namely backwashing, was found to be an effective strategy to control biomass growth [1].

In practice most off gas contaminated with VOCs have variable flowrates, unsteady loading, and contaminant compositions that limit the handling efficiency of the biofiltration [2]. The biofilter is also exposed to non-use periods as a shut down for equipment repair or during weekends and holidays. During such periods, the biofilter may be operated under the chemical starvation condition. It was reported that during the chemical starvation, the removal capacity of the biofilter decreased gradually [3]. In order to provide a constant high performance of the biofilter, the effect of non-use periods should be known to design the biofilter system and to identify operation criteria.

In this study, TBAB for VOC removal was operated under variable VOC loading. The experimental plan was designed to investigate the effect of non-use periods on the biofilter performance.

2. Experimental Methods

The experimental work was performed a laboratory scale trickle-bed biofilter for controlling VOC as a single contaminant. The model VOC was toluene (C_7H_8). A schematic diagram of the experimental system is presented in Figure 1. The biofilter is constructed of seven cylindrical glass sections with an internal diameter of 76 mm and a total length of 130 cm. Each section is equipped with a sampling port that extends to the centre of the column. The reactor is packed with pelletized diatomaceous earth biological support media (Celite® 6 mm R-635 Bio-Catalyst Carrier; Celite Corp., Lompoc, CA) to a depth of about 60 cm. The biofilter was initially seeded with an aerobic microbial culture, which was obtained from the secondary clarifier of an activated sludge system at a municipal wastewater treatment plant (Millcreek wastewater treatment plant, Cincinnati, OH). The biofilter has been maintained at a constant operating temperature of 20 °C in a constant temperature chamber and was operated in a co-current gas and liquid downward flow mode.

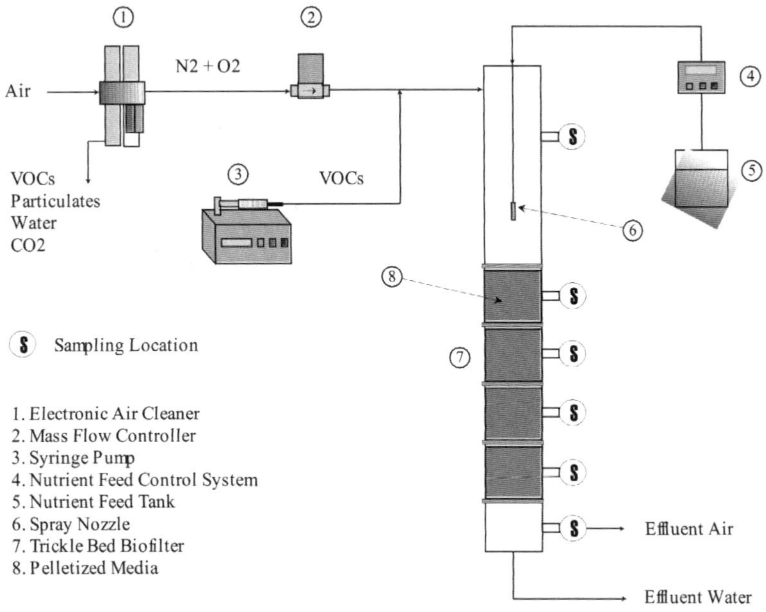

Figure 1. Schematic for the trickle bed air biofilter system

The air flow to the biofilter is set up at the rate of 3.6 L/min which is regulated by a mass flow controller. Liquid VOC was injected via a syringe pump into the air stream where it vaporized, and entered the biofilter through the topmost side port of the column.

The biofilter is equipped with an independent system for feeding 1.5 L/day of a buffered nutrient solution. The nutrient formulation for the biofilter is contains the same amount of nutrient-nitrogen and phosphorus ratio for a given VOC loading (a COD-to-nitrogen ratio of 50:1 and a nitrogen-to-phosphorous ratio of 4:1). As the sole source of nutrient-nitrogen, nitrate (NO_3-N) is used. The feed is sprayed as a fine mist onto the top of the medium bed through a spray nozzle.

Gas phase samples for VOC analysis were taken with gas-tight syringes. These VOC concentrations were measured by using a GC equipped with FID detector. Effluent gas phase sample for CO_2 analysis were also taken by using gas-tight syringes through sampling ports in the bifilters. A GC equipped with a thermal conductivity detector (TCD) was used for determining the CO_2 concentrations in the effluent gas phase. Liquid phase sample were analyzed for nitrate, total carbon (TC), inorganic carbon (IC), and volatile suspended solid (VSS) concentration.

Experiments were designed for maintaining consistently high performance of the biofilter. The biomass control strategy for this purpose involved in-situ upflow washing with nutrient solution, i.e., backwashing. The backwashing was conducted while the biofilter was off line by using 18 L of the buffered nutrient solution to induce full medium fluidization at about 50 % bed expansion for a defined time periods. The backwashing duration and frequency were initially set at 1 hour once per week (if necessary, 2 hours once per week were used).

Two different non-use periods were considered to mimic a shut down or break down of VOC source. The first is a period without chemical loading (namely, off chemical loading), which means pure air with nutrient flow through the biofilter. The second is a stagnant period (namely, off flow) which reflects no flow (chemicals, nutrient, and air) passing through the biofilter. The duration and frequency for each non-use period were two days per week for a period of three weeks at each VOC loading rate.

4. Experimental Results

4.1. Biofilter Performance

The biofilter was operated under the conditions as shown in Table 1.

Table 1. Experimental condition and strategy for toluene biofilteration

Operation Period	I	II	III	IV	V
Experimental Condition					
Inlet Concentration, ppmv	50	50	100	250	500
Loading rate, Kg COD/m^3·day	1.14	0.70	1.41	3.52	7.03
EBRT, min	0.76	1.23	1.23	1.23	1.23
Experimental Strategy (Operational periods in days)					
Backwashing (1 hour/week)	1 - 46	47 - 51	92 - 114	154 - 191	238 - 267
Non-use period					
Off chemical loading (2day/week)	-	52 – 71	115 - 133	192 - 210	-
Off flow (2day/week)	-	72 – 91	134 - 153	211 - 237	-

The first operation period (I) was considered as the acclimation period. The second operation period (II) consisted of the backwashing strategy as biomass control and two different non-use period strategies. So did the third (III) and the fourth (IV). During the fifth operation period (V), the overall toluene removal efficiency had not reached above 99 % under the backwashing strategy. The biofilter performance with respect to toluene removal is shown in Figure 2. For a VOC loading of 0.70 kg COD/m^3·day, the overall toluene removal efficiency reached over 99 % range for all experimental strategies. When the VOC loading was increased to 1.41 kg COD/m^3·day, a period of 10 days was required to acclimate the biofilter. After that, the toluene biofilter was maintained at the 99 % removal efficiency. During the non-use period strategy for VOC loadings of 0.70 and 1.41 kg COD/m^3·day, the 99 % removal efficiency was still maintained without backwashing strategy. For a VOC loading of 3.52 kg COD/m^3·day, a period of 15 days was demanded for acclimating the biofilter. The biofilter had been maintained at the 99 % removal efficiency. During the non-use period strategy for VOC loading of 3.52 kg COD/m^3·day, the removal efficiency dropped occasionally below 90 % and demanded the backwashing strategy as biomass control. It is interesting to note that the behavior of the biofilter depended on the biomass control because the biofilter performance decreased substantially with build up of back pressure due to accumulation of excess biomass [4-6].

A generalized summary of biofilter performance, at stable conditions, with respect to the experimental plans for different VOC inlet concentration is presented in Figure 3. It is worthwhile to note that the non-use period strategy provided similar biofilter performance as that with the backwashing strategy at 0.70 and 1.41 kg COD/m^3·day. Other study found that the biofilter under the controlled periodical operating condition can be operated without any contaminant breakthough [7]. Therefore, the non-use period can be considered as another means of the biomass control.

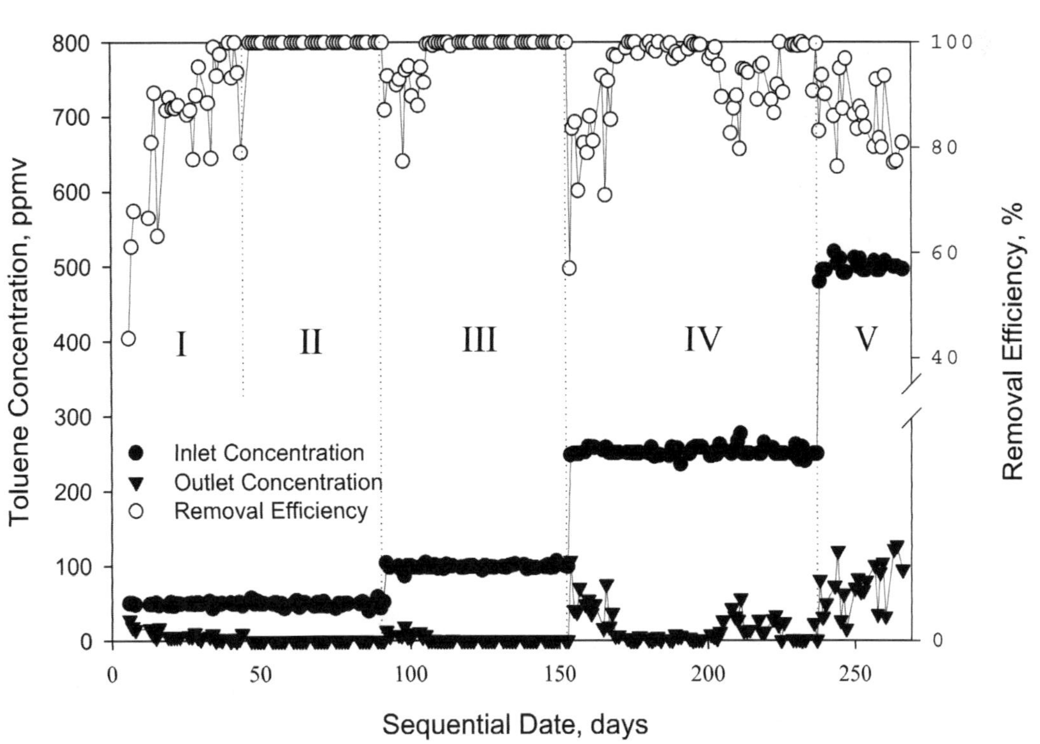

Figure 2. Biofilter performance with respect to toluene removal

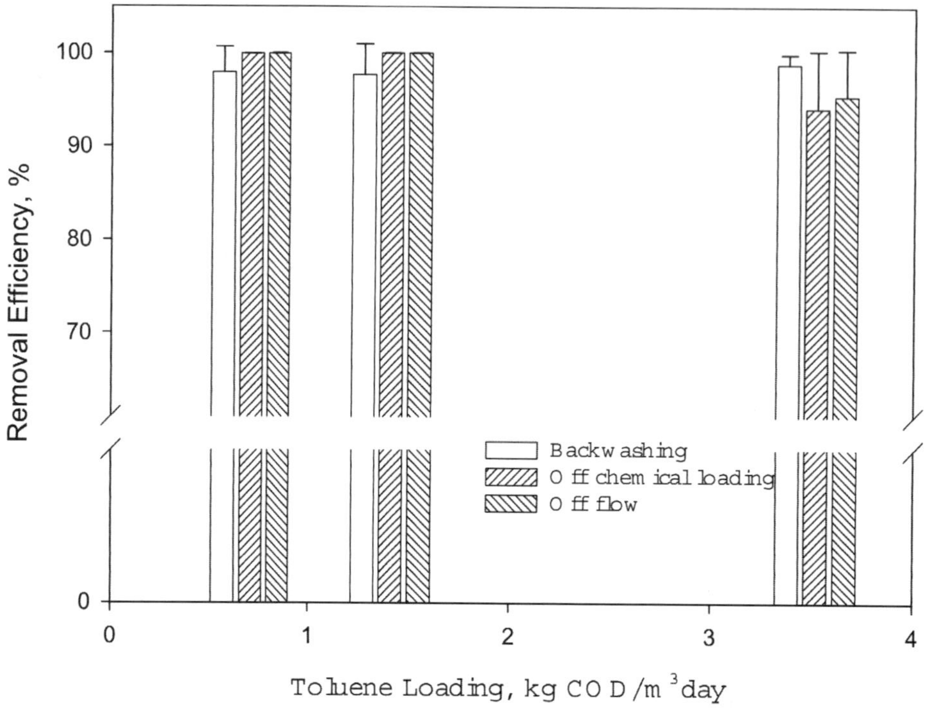

Figure 3. Biofilter performance with respect to experimental strategies

4.2. Biofilter Response after Backwashing and Non-use Period

The biofilter performance after backwashing and non-use periods was conducted by collecting effluent samples at prescheduled time intervals to determine the time varying response of the biofilter. Figure 4 shows the biofilter performance with time after the restart-up following the backwashing and the non-use period. The following observation can be deduced from Figure 4.

1. For the backwashing strategy, the overall removal efficiency was initially below the 30 % range but it increased over the 90 % range within 600 min at all inlet concentrations. The biofilter performance immediately after backwashing was unexpectedly poor due to the loss of active biomass.
2. For the two different non-use strategies, the biofilter initially had relatively high removal efficiency as compared with that for the backwashing strategy. The active biomass played an important role in this effect without the loss of active biomass.
3. At 250 ppmv of inlet concentration, higher removal efficiency had been observed in the initial time after the restart-up following the non-use period. It is speculated that initially, VOC adsorption on biofilm might occur because more biomass in the biofilter could build up at high VOC inlet concentration.

These observations suggested that the preservation of biomass may reduce the period of reacclimation of the biofilter performance after the non-use period. It also indicates that the biofiler performance following non-use periods is a strong function of the active biomass of the system. However, high biomass can hinder high performance of the biofilter due to increasing back pressure.

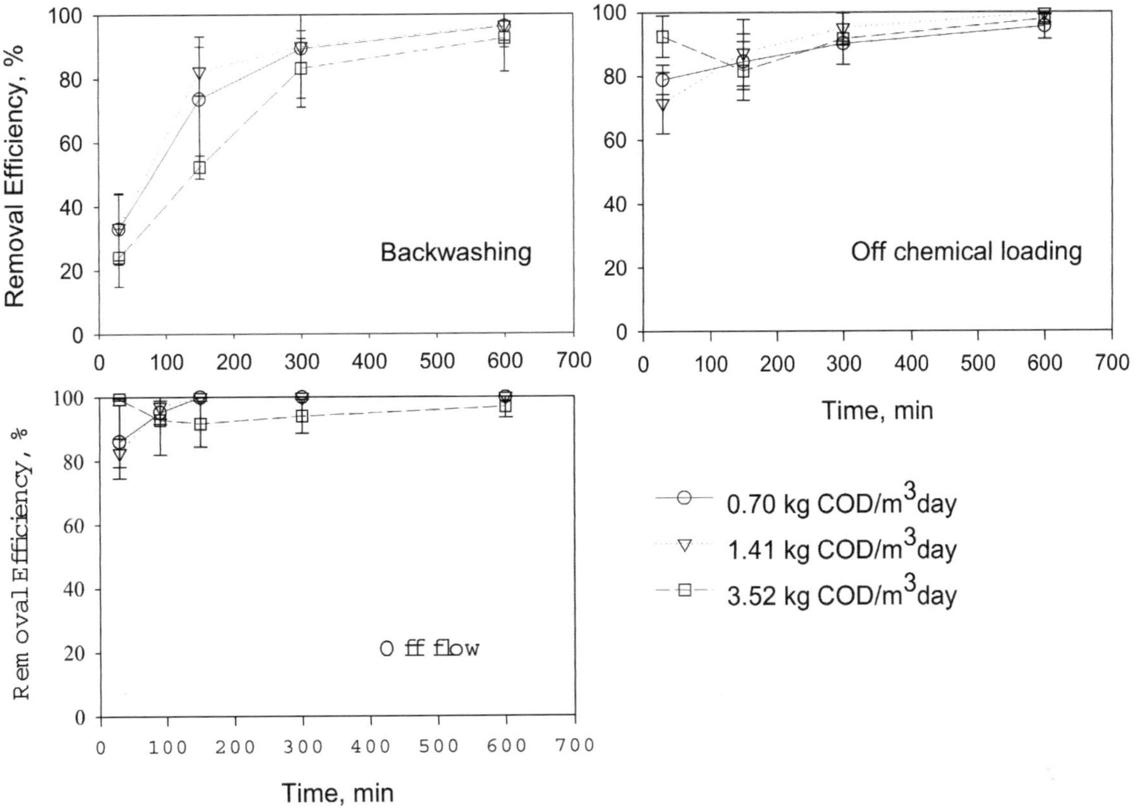

Figure 4. Biofilter performance after the restart-up following the experiment strategies at different VOC inlet concentration

4.3. Kinetic Analysis

Kinetic analyses were based on assuming first order removal rates as a function of depth in the biofilter for each experiment strategies. In order to avoid misinterpretation of the data due to the possibility of degradation occurring above the media bed along the reactor freeboard walls at top of the biofilter and in the bottom disengagement chamber used for separation between water and air, data obtained from sampling ports within the media were used for this purpose. The obtained results are shown in Table 2. It is seen that the removal rate was reduced with increase of VOC loading rate for all experimental strategies. At VOC loadings of 0.70 and 1.41 kg COD/ m^3·day, the non-use period strategy provided higher removal rate than that of the backwashing strategy, while at VOC loading of 3.52 kg COD/m^3·day, the backwashing strategy had slightly high removal rate. The experimental results show that the active biomass may play an important role in this effect as discussed previously.

Table 2. The reaction rate as a function of depth in the biofilter

	The first-order removal constant, sec^{-1}		
Loading rate, kg COD/m^3·day	Backwashing	Off chemical loading	Off flow
0.70	0.035	0.040	0.049
1.41	0.030	0.038	0.037
3.52	0.026	0.021	0.024
7.03	0.011	-	-

5. Conclusions

A laboratory scale trickle bed air biofilter was evaluated in this study for the treatment of air contaminated with toluene. The main focus was to study the effect of non-use periods on biofilter behavior for different VOC inlet loading and concentration. The following points were revealed:

1. High performance of the biofilter had been observed for the backwashing strategy and the non-use period strategy up to VOC loading of 3.52 kg COD/m^3·day corresponding to an inlet concentration of 250 ppmv as toluene.
2. The non-use period can be considered as another means of the biomass control for low VOC loadings (0.70 and 1.41 kg COD/m^3·day) but, at high VOC loading (3.52 kg COD/m^3·day), the active biomass control such as the backwashing is demanded.
3. There was little difference of biofilter performance between the two different non-use period strategies.
4. The biofilter response after restart-up following non-use periods was different with that after the backwashing. The biofilter performance after restart-up following non-use periods is a strong function of the active biomass in the system.
5. As the VOC loading rate was increased, much longer reacclimation period was necessary for biofilter performance to reach the 99 % removal range for both the backwashing and the non-use period.

Acknowledgements

The authors are pleased to acknowledge the financial support of the National Science Foundation under award number BES 0229135.

Reference

1. Smith, F. L.; Sorial, G. A.; Suidan, M. T.; Pandit, A.; Biswas, P.; Brenner, R. C. *J. Air Waste Manage. Assoc.* **1998**, *48*, 627-636.
2. Fitch, M. W.; England, E.; Zhang, B. *J. Air Waste Manage. Assoc.* **2002**, *52*, 1288-1297.
3. Cox, H. H. J.; Deshusses, M. A. *Environ. Sci. Technol.* **2002**, *36*, 3069-3073.
4. Smith, F. L.; Sorial, G. A.; Suidan, M. T.; Breen, A. W.; Biswas, P.; Brenner, R. C. *Environ. Sci. Technol.* **1996**, *30*, 1744-1751.
5. Sorial, G. A.; Smith, F. L.; Suidan, M. T.; Biswas, P.; Brenner, R. C. *J. Air Waste Manage. Assoc.* **1995**, *45*, 801-810.
6. Jorio, H.; Bibeau, L.; Heitz, M. *Environ. Sci. Technol.* **2000**, *34*, 1764-1771.
7. Irvine, R. L.; Moe, W. M. *Water Sci. Technol.* **2001**, *43*, 231-239.

A Novel Recycling Process of Waste Cement

Akihiro Yamasaki[1], *Atsushi IIzuka*[2], *Yasuro Katsuyama*[2], *Minoru Fujii*[3], *Yukio Yanagisawa*[2]

1. *Institute for Environmental Management Technology, National Institute of Advanced Industrial Science and Technology, 16-1 Onogawa, Tsukuba, 305-8569, Japan*
2. *Institute of Environmental Studies, Graduate School of Frontier Science, University of Tokyo, 7-3-1 Hongo, Bunkyo-ku, Tokyo, 113-8656, Japan*
3. *National Institute of Environmental Studies, 16-2 Onogawa, Tsukuba, 305-8506, Japan*

Abstract

A novel recycling process of waste cement by carbonation reaction with pressurized CO_2 is proposed. The proposed process consists of two main reactions. The first is the extraction of calcium ions from waste cement particles in water slurry by pressurized carbon dioxide (several MPa pressure). The second is the precipitation of calcium carbonate ($CaCO_3$) by reducing the pressure of CO_2. The product, $CaCO_3$ can be recycled as a raw material for cement production. In this process, the same amount of CO_2 emission is considered to be reduced because the net amount of virgin limestone mined can be reduced. Experimental studies were carried out in a laboratory-scale experimental apparatus to examine the feasibility of the proposed process. It was found that the extraction process can be proceed with relatively high rates in the initial 10 minutes of the reaction time, where the concentration of calcium ions exceeded thermodynamic solubility (supersaturation). The dependencies of the extraction reaction rate on the ratio of waste cement to water, CO_2 pressure, and the particle size of the raw waste cement particles were investigated experimentally with a laboratory-scale setup. Calcium carbonate with about 97 % purity was obtained in the precipitation process. A process simulation was conducted based on the experimental results, and energy consumption and the cost for a model process were estimated. The cost for the recycling of 1 metric-ton of $CaCO_3$ is about US$ 16.6, which is much lower than the market price of calcium carbonate.

1. INTRODUCTION

Waste cement is a calcium-rich waste product containing calcium in the form of calcium silicate hydrate (such as C_3S; $3CaO \cdot 2SiO_2 \cdot 3H_2O$) and calcium hydroxide ($Ca(OH)_2$). Waste cement is produced in the process that recovers aggregates from waste concrete. While recycling and reuse of the recovered aggregates is an established process, reuse of waste cement has been limited to use as a roadbed material, and no effective utilization has yet been realized. Because of a continuous increase in the supply of waste concrete from dismantled buildings, a long-term copious and stable supply of waste cement is expected in the future. One potential use of waste cement is use as a calcium source for cement production. Carbonation reaction of waste cement (represented by $3CaO \cdot 2SiO_2 \cdot 3H_2O$ and $Ca(OH)_2$) would proceed in the following schemes,

$$3CaO \cdot 2SiO_2 \cdot 3H_2O + 3CO_2 + 3H_2O \rightarrow 3CaCO_3\downarrow + 2SiO_2 + 6H_2O \quad (1)$$

$$Ca(OH)_2 + CO_2 + H_2O \rightarrow CaCO_3\downarrow + 2H_2O \quad (2)$$

However, the above reactions would proceed with relatively slow reaction rate under the atmospheric condition, *i.e.*, low pressure of CO_2. In reality, the carbonation of cement in the buildings could proceed with very limited rate. For a practical recycling process of waste cement streams, an acceleration of the carbonation process is essential. For the purpose of the acceleration of the carbonation reaction of the waste cement, we have proposed a new type of reaction scheme consisting of two steps of the extraction of calcium ions from the waste cement under high-pressure of CO_2 and the precipitation of calcium carbonate under reduced pressure of CO_2. In this paper, the feasibility of the proposed process was examined in terms of the energy consumption and the cost based on the experimental results in a laboratory scale.

2. OUTLINE OF THE RECYCLING PROCESS

The first step is an extraction reaction of calcium ions by a pressurized CO_2 from the waste cement particles in the form of water slurry. The scheme of the first step can be described as,

(1) Extraction step

Dissolution of CO_2 in water

$$CO_2 + H_2O \rightarrow H_2CO_3 \quad (3)$$

$$H_2CO_3 \rightarrow H^+ + HCO_3^- \quad (4)$$

$$HCO_3^- \rightarrow H^+ + CO_3^{2-} \quad (5)$$

Extraction reaction of waste cement with protons

$$3CaO \cdot 2SiO_2 \cdot 3H_2O + 6CO_2 + H_2O \rightarrow 3Ca^{2+} + 2SiO_2 + 6H_2O + HCO_3^- + CO_3^{2-}$$

(6)

$$Ca(OH)_2 + 2CO_2 + H_2O \rightarrow Ca^{2+} + 2H_2O$$

(7)

After separation of the extracted solution from the slurry, calcium carbonate is allowed to be precipitated from the solution by reducing the CO_2 pressure. Thus, the second step is the precipitation reaction of calcium carbonate from the extracted solution. The reaction scheme of the second step is described as follows,

(2) Precipitation step

$$Ca^{2+} + CO_3^{2-} \rightarrow CaCO_3\downarrow \quad (8)$$

The advantage of applying pressurized CO_2 in the extraction step is both in accelerating the extraction reaction rate and in enhancing the extraction ratio of calcium carbonate from the waste cement. Since the precipitation reaction is conducted under the reduced pressure conditions, the difference in the solubility of calcium ions between the two conditions of the extraction step and the precipitation step will determine the conversion of the calcium carbonate as the product. Since the solubility of calcium ions highly depends on the CO_2 pressure, the larger pressure difference between the two steps will increase the conversion of the calcium carbonate.

The use of CO_2 as an extraction media will lead the reduction of CO_2 emission, which could recognized as a sequestration option of carbon to prevent global warming. In this case, the use of CO_2 produced from a point source such as thermal power plant is considered. The CO_2 will be fixed in the form of carbonate in the recycling process, but the fixed CO_2 will be reemitted in the cement production reaction,

$$CaCO_3 \rightarrow CaO + CO_2$$

(9)

In such a case, the net reduction of CO_2 can be achieved because the emission in the cement production process can be reduced by the amount of CO_2 that would potentially be emitted from virgin limestone.

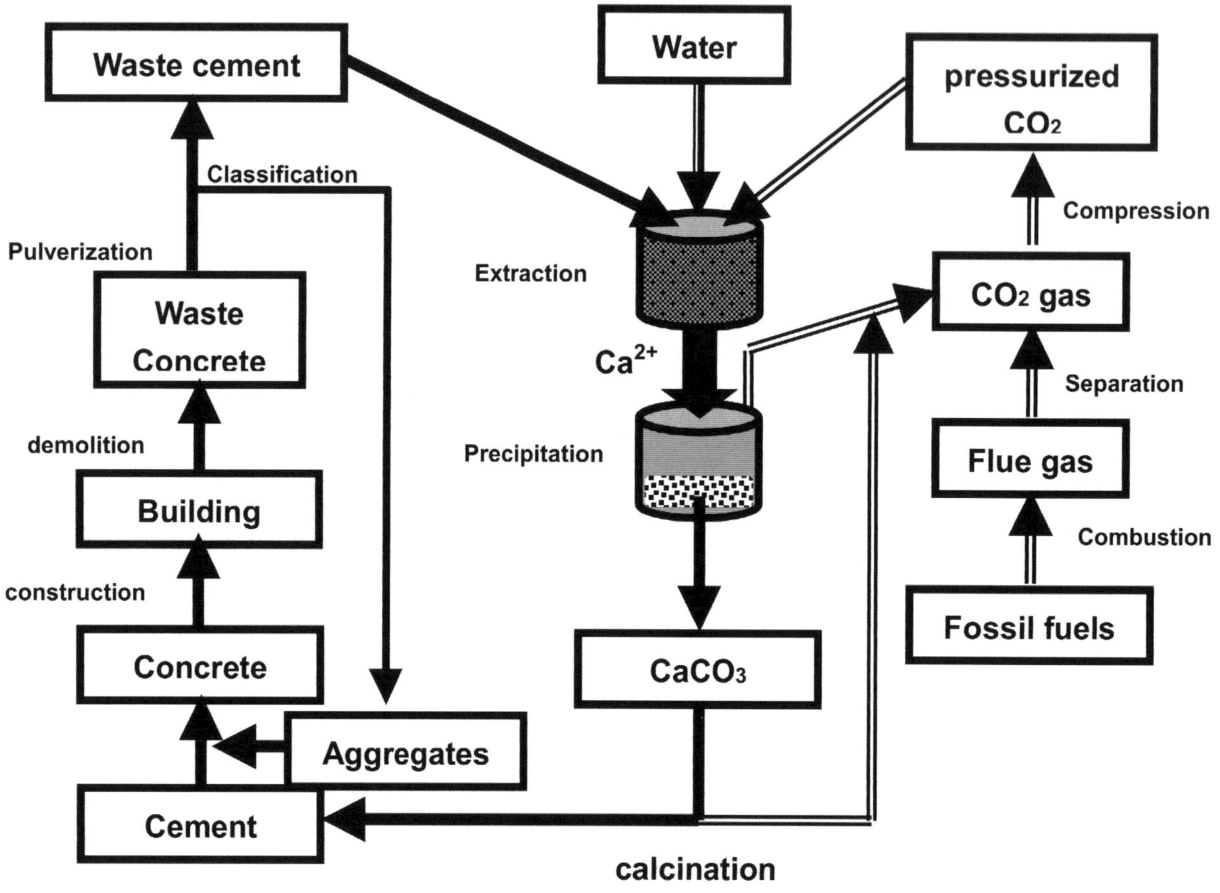

Figure 1 Outline of the recycling process of waste cement

Process flow diagram of the proposed process is shown in Figure 1. Waste concrete discharged from dismantled buildings is first pulverized, and classified into aggregates and waste cement particles. The waste cement particles are fed into a high-pressure reactor vessel, to extract calcium ions with pressurized CO_2 and water. The CO_2 is captured and separated from concentrated emission sources such as thermal power plants. In the extraction vessel, the extraction of calcium ions from waste cement (reactions (6) and (7)) would take place with stirring. After a certain period of time, the slurry is filtered and the solution phase is separated. Then, the solution is left under a reduced pressure of CO_2 in the settling bath to allow precipitation of $CaCO_3$ to take place. The $CaCO_3$ product can be used as a raw material in the cement production industry. The remaining

CO_2 will be recycled to the extraction process after re-pressurization. Although the above reaction scheme would proceed spontaneously from the thermodynamic point of view, kinetic data are necessary to design and evaluate the process.

4. EXPERIMENTAL

4.1 Characterization of the waste cement sample

The waste cement produced in a real recycling process was used, which was kindly supplied by Tateishi construction corp. The waste cement is in the form of fine particles with the size of 100 ~ 200 •m, with a peak at about 25 ~ 40 •m (area based), or about 80 •m (volume based). The composition of the waste cement was analyzed by a fluorescent X-ray method (JEOL, JSX-3220), and the weight fraction of calcium was determined to be about 27.3 %. The degree of carbonation of the waste cement was determined by a thermo-gravimetric method with a differential thermal analyzer (ULVAC, TGD-9600), and found about 11 % of the total calcium in the waste cement was carbonated. Thus, calcium content available for 24.3 weight % of the waste cement was available for the CO_2 treatment in the present experiments.

4.2 Calcium extraction experiments

4.2.1 Experimental apparatus and procedure

A stirring vessel reactor was used for the calcium extraction experiments. The vessel is made of HASTELLOY® of which the inner volume is 500 mL. A known amount of the waste cement particles was introduced into the reactor under atmospheric conditions, and mixed with a given amount of deionized water. Flow of gaseous CO_2 was continuously supplied to the reactor of which the pressure is controlled by a backpressure-controlling valve installed at the downstream side of the reactor. The reactor was immersed in a constant temperature bath, and the reaction temperature is controlled with an accuracy of ± 1 °C. The stirring in the reactor was carried out using a two-wing paddle type fin, and the stirring rate was controlled at rates in the range of 0 ~ 1000 rpm.

The contents in the reactor were sampled through a sintered-metal filter with 5 μm mesh at given intervals during the extraction experiments. The filtered solution was then depressurized to atmospheric pressure, and the concentration of calcium ions was determined by inductively coupled plasma - atomic emission spectrometry (ICP-AES, Hitachi, P- 4010). The chemical composition remained particles after evaporation of water from the sampled solution were identified, using X-ray diffraction analysis (Rigaku, Mini Flex) and fluorescent X-ray analysis (JEOL, JSX-3220), and a thermo-gravimetric method with a differential thermal analyzer (ULVAC, TGD-9600).

4.2.2 Experimental results

The X-ray diffraction pattern of the particles remaining after evaporation of water from the sampled solution was similar to that of calcite. Fluorescent X-ray analysis demonstrated that the calcium fraction of the metal content in the remaining particles was larger than 97 %. These results indicate that high purity $CaCO_3$ (calcite) can be obtained from the waste cement particles by the extraction procedures used in these experiments.

Figure 2 shows the time course of the calcium concentration in the extracted solution for various initial amounts of waste concrete. Common experimental conditions were, stirring speed = 900 rpm, temperature = 323 K, p_{CO2} = 3.0 MPa.

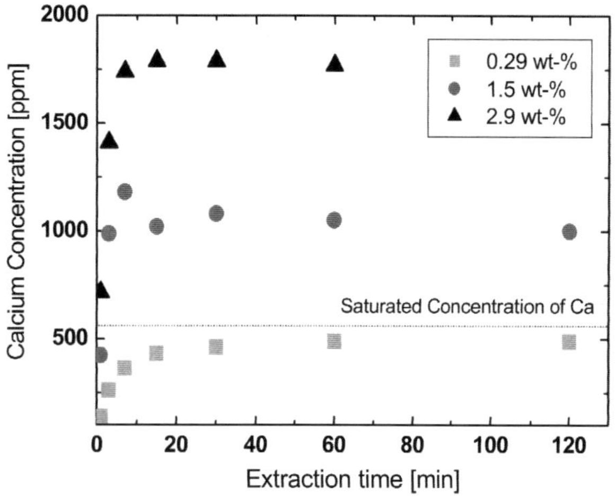

Figure 2 Time course of the calcium concentration in the extracted solution for various cement/water ratios.

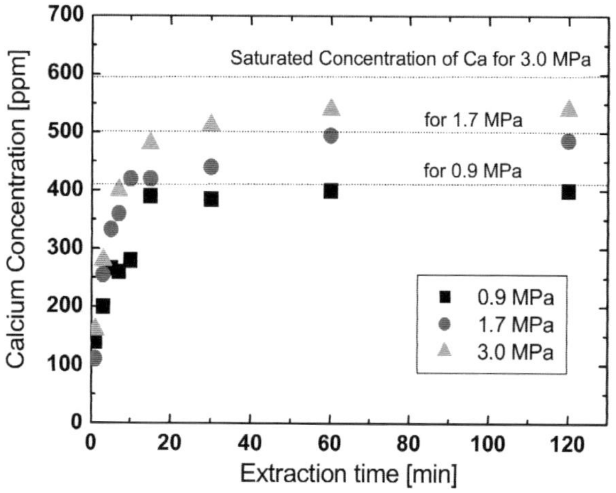

Figure 3 Time course of the calcium concentration in the extracted solution for CO_2

The calcium concentration increased rapidly at the initial stage of the reaction up to about 10 min, and after that, the concentration leveled off or decreased slightly. The dotted line in the figure represents the solubility of calcium ions. The sampled solution was supersaturated with calcium ions even at the early stage of the extraction reaction, except in the case with the smallest ratio of waste cement / water, *i.e.*, weight fraction of waste cement is 0.29 wt%. The occurrence of the supersaturation could be explained in terms of the rate balance of the extraction and the precipitation. In general, the extraction reaction would proceed much faster than the precipitation reaction, and consequently, the concentration will exceed the thermodynamic limitation at the initial stage. At the later stage of the reaction, however, the precipitation occurs gradually, and its rate is balanced by the extraction rate, which would reduce the apparent extraction rate. For the case of the small ratio of waste cement / water, the extraction rate would be balanced by the deposition rate at an earlier stage of the reaction experiment, and the concentration of calcium ions would not exceed that given by the solubility line.

The extraction ratio of calcium after 3 min extraction time was about 36 % for the run with the smallest weight fraction of waste cement /water (0.29 wt%), and 18 % for the run with the largest weight fraction (2.9 wt%). Note the extraction ratio is defined as the molar ratio of the extracted calcium to the calcium in the feed cement. Thus, the extraction ratio could be improved in the case of smaller ratio of cement to water in the feed.

Figure 3 shows the influence of the CO_2 pressure on the concentration change of calcium ions. Experimental conditions were: stirring rate = 900 rpm, temperature = 323 K, initial amount of waste cement = 1.0 g in 350 g of water (weight fraction of waste cement is 0.29 wt%). The initial extraction rate was found to increase with an increase in the CO_2 pressure. The increase in the initial extraction rate with increasing CO_2 pressure could be attributed to the increase in the solubility of calcium ions, and consequently, an increase extraction rate.

The initial extraction ratio was found also increased by the decrease in the average particle size of the feed cement particles. This is mainly due to the larger specific surface are for the smaller particles would increase the apparent extraction rate.

5. PROCESS DESIGN AND EVALUATION

Since the present process is partially aiming the reduction of CO_2, it is assumed that the CO_2 necessary for the treatment of waste cement is supplied from a thermal power plant, and the process flow is designed on the basis of the CO_2 flow rate from the power plant. The net amount of CO_2 emission from the power plant is set at 0.0174 t / s [1] assuming a coal-fired thermal power plant with 40 % efficiency. This flow rate will produce $CaCO_3$ at the rate of 0.0395 t/s. Since the weight fraction of active calcium in the present waste cement sample is about 24.3 %, 0.0716 t / year of waste cement could be treated by the CO_2 emitted from the power plant, assuming 100 % conversion.

As a pre-treatment of the waste cement, the waste cement particles will be pulverized from 5 cm (typical recycling case) to 200 •m (for the experimental condition in the present study) to increase the extraction rate. The pulverization energy was calculated based on Bond's equation. The CO_2 will be captured from the flue gas of the thermal power plant by an amine absorption/desorption process, and the energy for the capturing CO_2 is cited from a literature [1].

The CO_2 pressure for the extraction reaction is 3.0 MPa of pure CO_2, and the captured CO_2 will be pressurized from 0.1 MPa to 3.0 MPa by an adiabatic compression process with 90 % efficiency before introduced to the extraction reactor. The extraction reaction is assumed to be carried out with 20 identical reactors, and the reactor volume was calculated assuming the retention time at 3.0 min. or volume required was determined from the CO_2 emission rate and the retention time in the extraction reactor. The stirring power was estimated by adjusting the retention time and the power consumption for stirring per unit volume of the content relative to those of the extraction reactor used in the present experiments. Only the power consumption for the stirring was considered for the required energy estimation in the extraction reaction. Waste heat from the thermal power plant is used for heating the reactant up to the extraction temperature.

Energy requirement for the precipitation process was not considered because it is negligible compared with those for the preceding extraction and pre-treatment processes.

The power consumption for the pressurization of CO_2 will consume the largest portion of energy in any cases. The process was optimized in terms of the energy consumption, and the minimum energy consumption among the cases studied was about 25.9 MW, where the weight fraction of waste cement is 2.9 wt% (water balance), p_{CO2} = 3.0 MPa, and temperature = 323 K. The breakdown of the energy consumption for the producing calcium carbonate from the waste cement for an optimum case is summarized in Table 1. The cost for recycling of $CaCO_3$ is as large as 16.56 US$/metric ton, which is much lower than the market price of $CaCO_3$ (about US$ 200 for low-purity $CaCO_3$, about US $400 for high-purity $CaCO_3$). A rough estimation of the capital cost of treatment plant would double the cost, but still the present process is competitive with the conventional method. In addition, about 0.44 ton of CO_2 could be fixed per ton $CaCO_3$ production in the treatment process, which could further reduce the cost of production if carbon tax is introduced.

Table 1 Breakdown of the power consumption for the production of $CaCO_3$

	CO_2 Separation	Pressurization	Stirring	Pulverization	Total
Power Consumption [MW]	5.5	13.7	0.32	6.4	25.9
Energy Consumption [kWh / t]	39.1	97.9	2.3	45.8	185.1
Cost [USD / t]	3.48	8.76	0.24	4.08	16.56

References

1. Halmann MM., Steinberg, M. Greenhouse Gas Carbon Dioxide Mitigation" Lewis Publishers Inc., (1998).

A Life Cycle Assessment on Alternative Options for Recycling of Plastic Packaging Wastes.

Floriana PERUGINI, Umberto ARENA and Maria Laura MASTELLONE

Dept. of Environmental Sciences - University of Naples II, Via Vivaldi, 43 – 81100 Caserta (Italy)
e-mail:floriana.perugini@unina2.it;umberto.arena@unina2.it; mlaura.mastellone@unina2.it

INTRODUCTION

A Life Cycle Assessment (LCA) methodology is generally considered one of the best environmental management tools that can be used to obtain a Progress Integration policy for waste recycle and management (McDougall et al., 2001; Arena et al., 2003a and 2003b). European Community officially recognized (Council Directive 91/61/EC/1996) that the LCA tool could play a crucial role in an objective selection of process. It is able to evaluate and compare alternative eco-performances by considering the environment as a whole, including direct releases, consumption of raw materials and waste disposal. LCA offers a systematic way to incorporate the entire material and energy supply chain into strategic planning and policy development (Azapagic, 1999).

The overall goal of the study was the whole Italian system for recycling of household plastic packaging wastes. The phases of different management options were individually analyzed and quantified in terms of energy and material consumptions as well as of emissions in the environment. More specifically, the study aimed to develop information able to evaluate the environmental performance of alternative plastic management options in Italy, to quantify the overall environmental performance of mechanical recycling of plastic containers and to compare it with that of conventional options of landfilling or incineration and with that of an innovative processes of feedstock recycling, the British Petroleum polymer cracking, obtained by means of a low-temperature fluidized bed pyrolysis.

THE APPROACH TO THE ANALYZED PLASTIC WASTE MANAGEMENT

The *overall goal* for the project is the whole Italian system for recycling of household plastic packaging wastes, in particular liquid containers mainly made of PET and PE. The *primary audience* for this effort is the Italian National Consortium for Packaging (Conai), which is interested to assess energetic and environmental profile of the current plastic waste management in comparison with some alternatives. The function of the system under study is to manage (and recycle) plastic containers for liquids. Therefore, the *functional unit* has been defined as the management of post-consumer PE and PET liquid containers (obtained by means of a mono-material collection), which leads to the production of 1kg of polymeric flakes. It is assumed that there is a market demand for all recycling products under investigation and that virgin and recycled material are equivalent for the market.

Four scenarios of management of plastic wastes were then defined:

Scenario I. **No recycling and landfill disposal** of all the collected plastic wastes;

Scenario II. **No recycling and** all the collected plastic wastes sent to **combustion with energy recovery**;

Scenario III. **Mechanical recycling** of all the collected plastic wastes and all the process wastes sent to incineration with energy recovery.

Scenario IV. **Mechanical recycling for PET plastic wastes and Low-Temperature Pyrolysis for polyolefin fraction** of plastic wastes.

It was assumed that the plastic container for liquids enters the *system boundaries* when it is delivered to a single material collection site, whether it is a residential kerbside, apartment collection site, or rural drop-off site. In agreement with similar studies (McDougall et al, 2001), all upstream life cycle activities (raw materials extraction, manufacturing, and use) are assumed to be held constant. Moreover, the scope of this study does not include general industrial and commercial waste not collected with the municipal solid waste. Finally, the study does not cover the economic aspects of waste management: no assessment will be made of the economic impact of the various scenarios. Figure 1 schematically describes the system boundaries for one of the considered scenarios.

The phase of Life Cycle Inventory over the system under study aimed at identifying and quantifying the environmental burdens crossing system boundaries: it resulted in a list of raw materials and energy inputs and of individual emissions to air and water and solid waste generation, the inventory table, that is reported in detail elsewhere (Fraunhofer-Institut, 2001, Arena et al., 2003a and 2003b). In particular, the *direct environmental burdens* (i.e. those associated with plastic waste recycling activities) were evaluated for each stage of the recycling chain as averaged from on site investigation data or official documents. The *indirect burdens* (i.e. those associated to the production, transport and use processes of all the needs, raw materials, energy, apparatus and other, for the carrying out the different plastic waste recycling activities) as well as the *avoided burdens* (i.e. those which have to be deducted from the total environmental impact account since related to products obtained from one of the processes and therefore not more to be generated by means of traditional activities) were instead calculated by means of the Boustead Ltd data bank. Figure 2 highlights the avoided burdens that characterize the four management scenarios.

THE LIFE CYCLE IMPACT ASSESSMENT RESULTS

It is generally acknowledged that the phase of methodological and scientific framework for life cycle impact assessing is still being developed (Clift et al., 2000), even though some relevant improvements have been obtained in the last years (SETAC, 2002). As a consequence, in this analysis, according to almost all of similar studies (Arena et al., 2003a and 2003b), the impact categories, bottom-up method was utilized.

The analytical comparison between the four management scenarios (Fig. 2) is described by the diagrams reported in the following with reference to some crucial impact categories (Figures 3-5). All of them refer to the functional unit, i.e. the basis for comparison of different management schemes. This latter has been defined as the management of plastic waste that leads to the production of 1kg of (recycled or virgin) PET flakes and of 0.39kg of (recycled or virgin) PE flakes. On the basis of previous definitions, the negative values in the figures indicate the predominance of avoided environmental burdens.

It appears that the mechanical recycling option (Scenario III) is always environmentally preferable, even though the energy recovery option also appear to guarantee a remarkable energy saving (Figure 3). The poor performance of landfilling option is clearly indicated by all the impact category indexes. The analysis of Figure 3A, in particular, suggests a not positive performance of the management scenario including mechanical recycling of PET and feedstock recycling of polyolefins. On the other hand, this option presents a large saving of not renewable resources as shown by Figure 3B and detailed by Figure 4. The latter highlights the positive role that the low-temperature pyrolysis process has in the saving of crude oil resource. The good performance of scenario including feedstock recycling is also indicated by Figure 5 that reports the amounts of GHG generated by the different scenarios: again both the recycling options present the best performance.

Table 1 summarizes the indicators of principal environmental impact categories, as evaluated for the four scenarios. The previous considerations are clearly summarized. In particular, the overall performance of plastics management scheme adopted in Italy is rather satisfactory having in almost all the cases the best environmental performance.

REFERENCES

Arena, U., M.L. Mastellone, F. Perugini, "Life Cycle Assessment of a Plastic Packaging Recycling System", *Int. J. LCA*, **8**(2):92-98 (2003a)

Arena, U., M.L. Mastellone, F. Perugini, "The Environmental Performance of Alternative Solid Waste Management Options: a Life Cycle Assessment Study", *Chem. Eng. J.*, **96**:207-222 (2003b)

Azapagic Adisa "Life cycle assessment and its application to process selection, design and optimisation" *Chem. Eng. J.,* **73**: 1-21

Clift, R., A. Doig, G. Finnveden, "The Application of Life Cycle Assessment to Integrated Waste Management. Part 1- Methodology", *Trans IchemE*, **78**(B)279-287 (2000)

Council Directive 91/61/EC 1996. Concerning Integrated Pollution Prevention and Control, Official Journal of the European Communities, No. L 257, 10 October, HMSO, London, 1996.

Fraunhofer-Institut, "Feedstock Recycling of Plastic Waste in the Polymer Cracking Process of the BP-consortium. LCA type analysis", Final Report for APME, June 2001

SETAC, "Life-Cycle Impact Assessment: Striving Towards Best Practice", Udo de Haes et al. (eds.), Setac Press, Brussels, ISBN 1-880611-54-6, 2002

McDougall, F.R., P. White, M. Franke and P. Hindle, *Integrated Solid Waste Management: a Life cycle Inventory*, 2nd ed., Blackwell Science (2001)

Table 1 The indicators of principal environmental impact categories, as evaluated for the four scenarios for plastic waste management. The bold types indicate the best environmental performance while the italic types indicate the worse.

Impact category	Scenario I	Scenario II	Scenario III	Scenario IV
Energy consumption, $MJ/kg_{\text{recycled or virgin PET}}$	*51.59*	6.45	**-5.41**	12.14
Crude oil consumption, $g/kg_{\text{recycled or virgin PET}}$	*1462*	995	-74.76	**-145.86**
Water consumption, $kg/kg_{\text{recycled or virgin PET}}$	*47.11*	45.92	**3.48**	14.06
CO_2-equivalent, $kg/kg_{\text{recycled or virgin PET}}$	5.3	*7.3*	**1.4**	1.7
Air emissions of organic compounds, $g/kg_{\text{recycled or virgin PET}}$	*26.8*	14.3	**-0.05**	1.42
Waste production, $kg/kg_{\text{recycled or virgin PET}}$	*2.49*	0.19	**0.09**	0.2

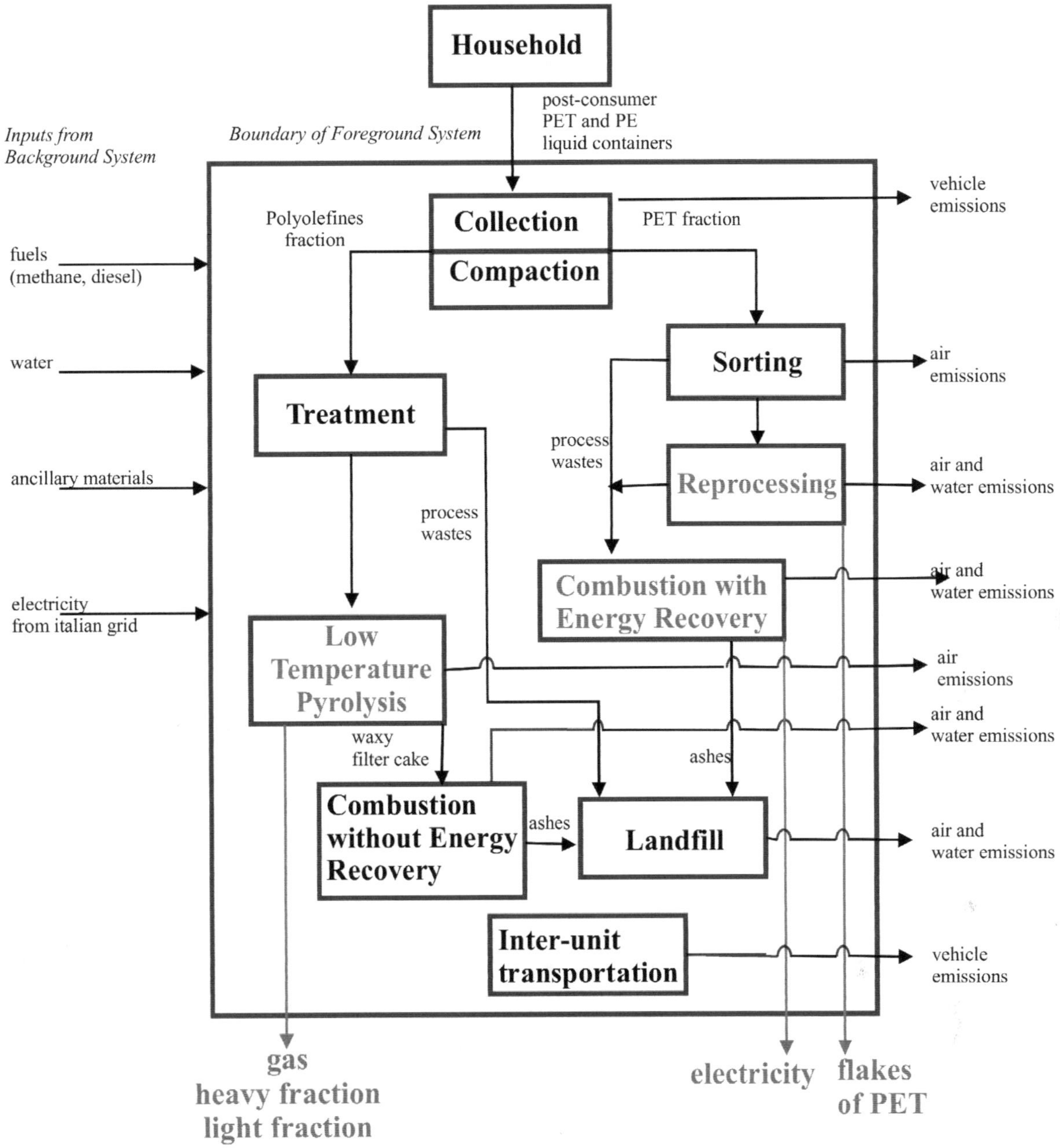

Fig. 1 System boundaries with the indication of typology of environmental burdens (scenario of mechanical recycling for the PET fraction and of Low-Temperature pyrolysis for the polyolefins fraction).

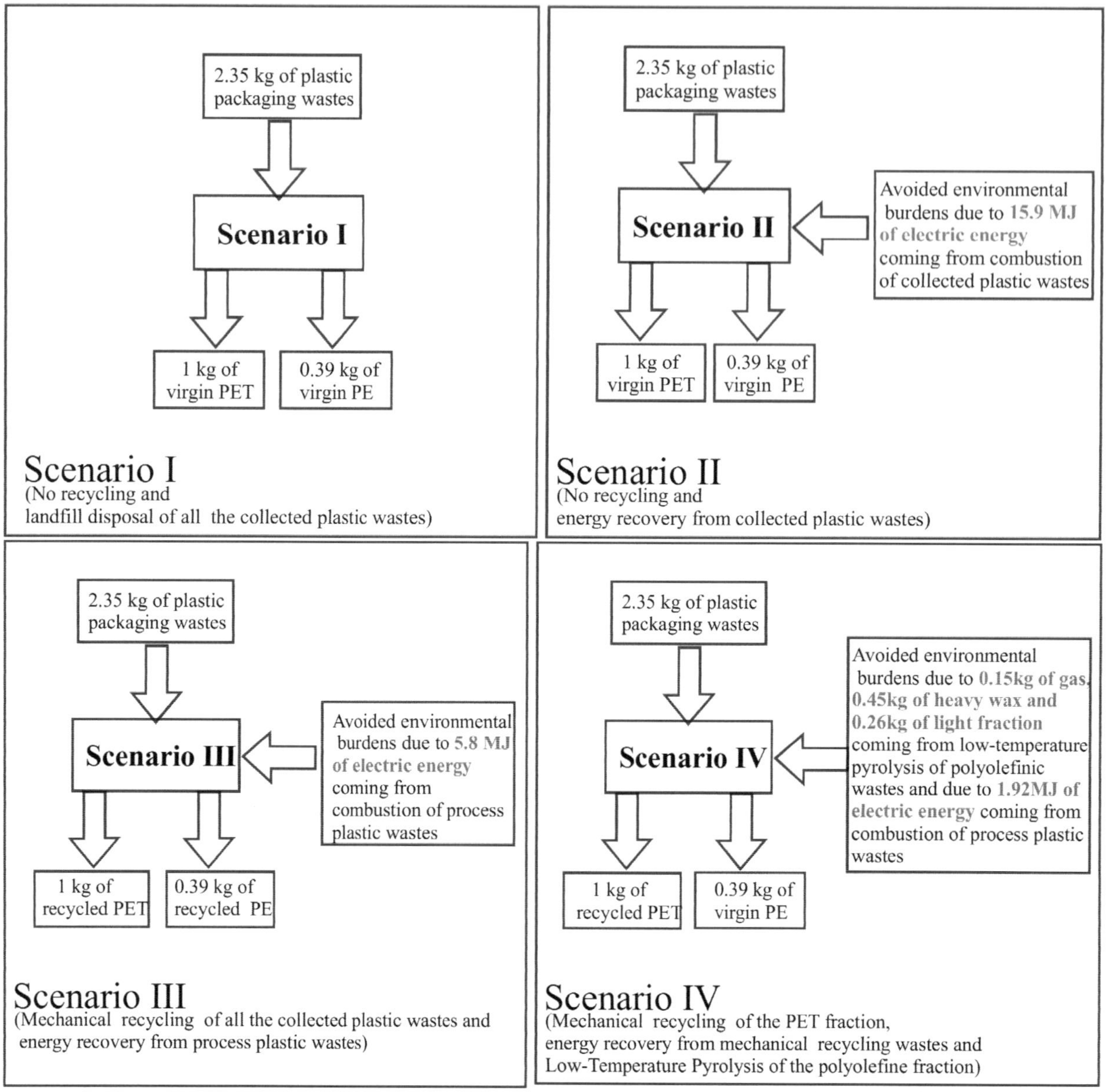

Fig. 2 Basket of products for the plastic waste management scenarios with the indication of avoided environmental burdens.

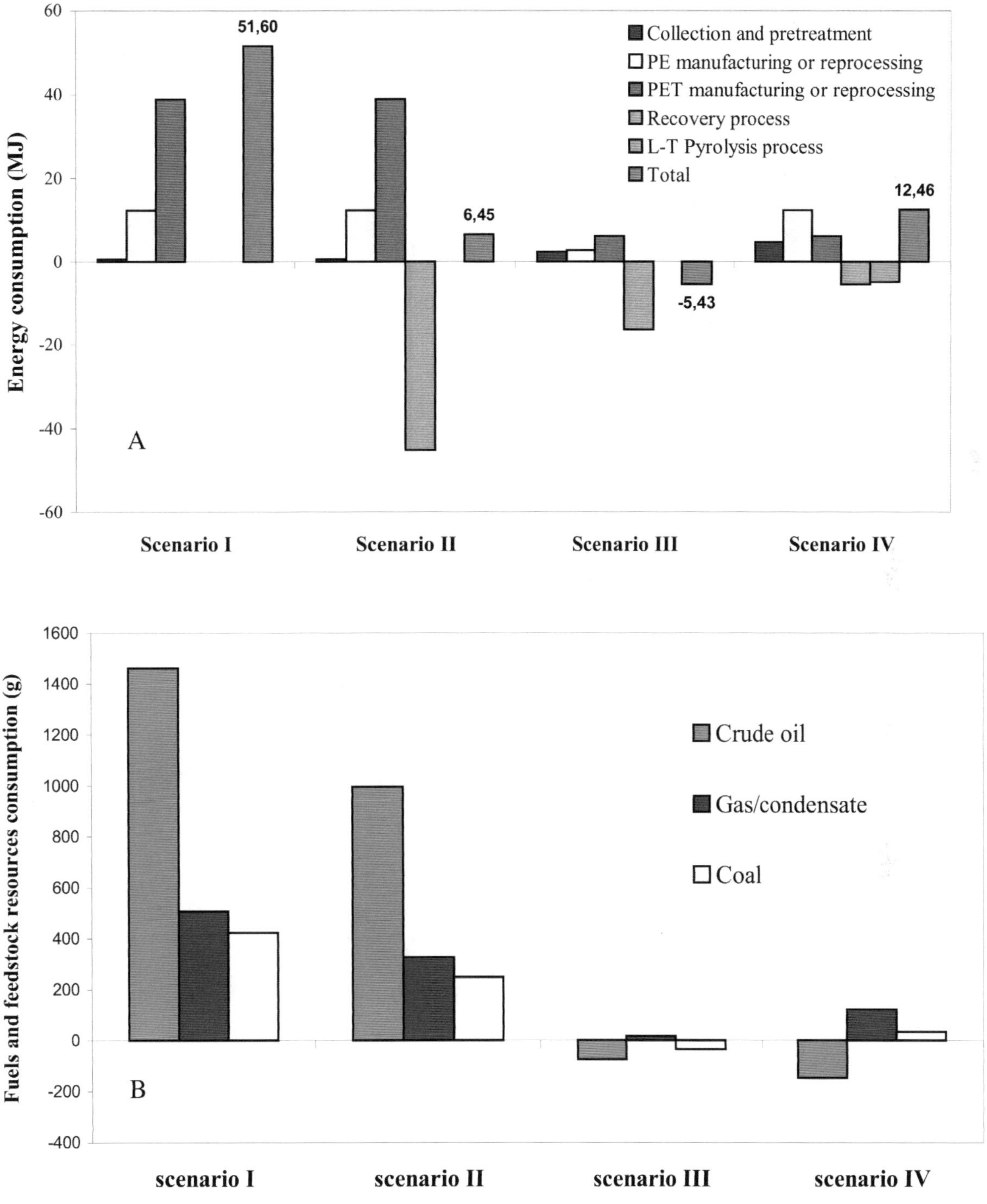

Fig. 3 The energy consumption related to each plastic waste management scenario with (A) the indication of contributions coming from the different stages, and (B) the fuels and feedstock resources consumption.

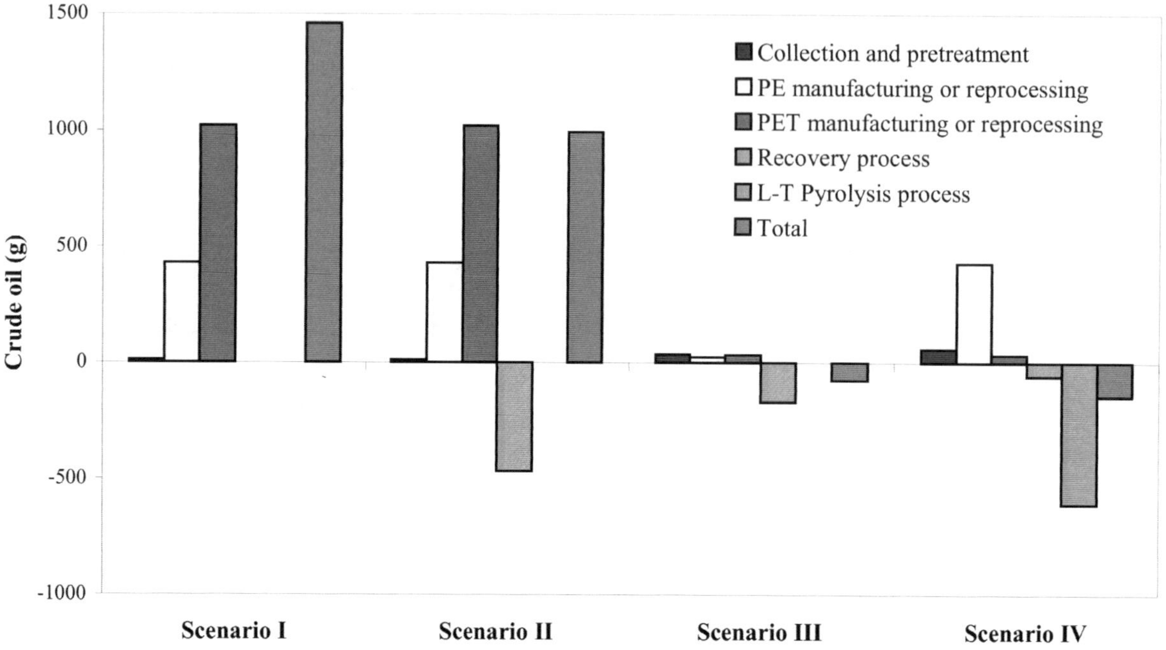

Fig. 4 The crude oil consumption related to each plastic waste management scenario with the indication of contributions coming from the different stages.

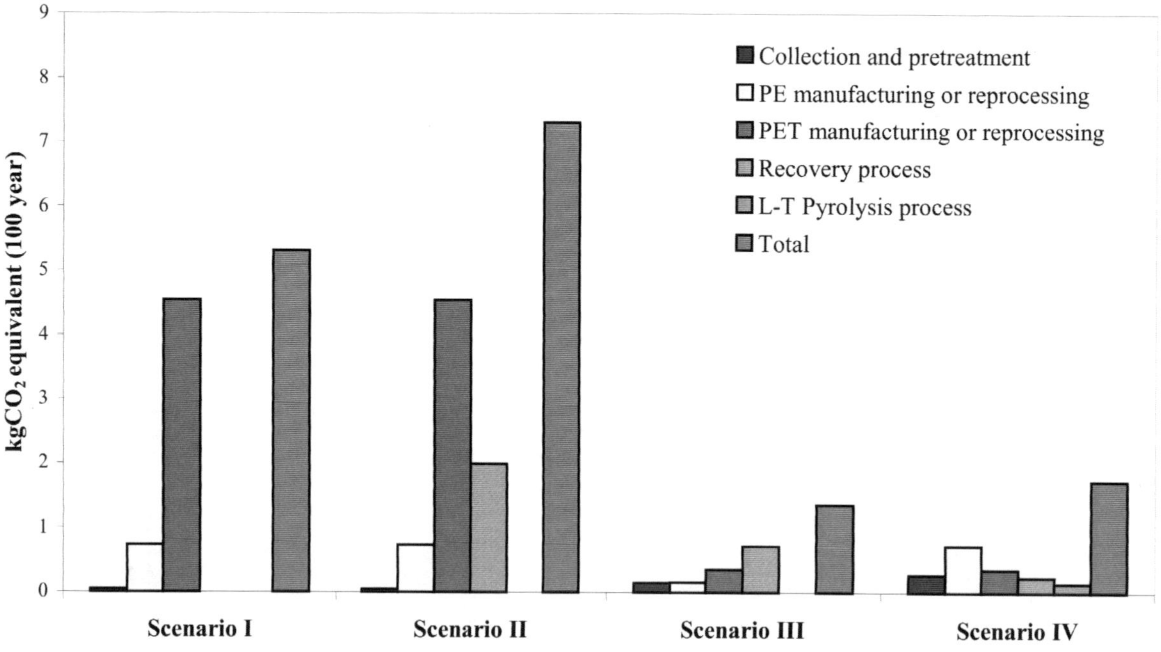

Fig. 5 The generation of green house gases (expressed as kilograms of CO2 equivalent) related to each plastic waste management scenario with the indication of contributions coming from the different stages.

TARGETING FOR ENERGY SAVING IN POLYCARBONATE PLANTS USING REAL-TIME ENERGY INTEGRATION

Noureldin, M.B., Hasan, A.N.,

Department of Materials and Process Engineering
The University of Waikato, Hamilton, New Zealand

ABSTRACT

The worldwide increase in energy demands, together with the uncertainty of future energy costs, make energy conservation a major element in controlling operating costs of chemical processes. Energy usage in chemical facilities represents a major cost center and process owners must constantly monitor its consumption to enhance their profitability. Keeping industrial energy consumption bill under control not only enhances economic competitiveness but also minimizes the depletion of unsustainable conventional energy resources, and protects the environment.

Such needs have motivated researchers to investigate more aggressive energy recovery schemes. These efforts do not see widespread applications in industry since energy is still considered as a process constraint and rarely being seen as one of the process objectives. Real-time operation optimization of chemical plants, for the sake of process energy saving, is a rare task and always lack justification. Two main challenges are facing the industrial community. The first one is related to the cost of dealing with the combinatorial nature of the energy problem in complex industrial facilities, especially the on-line problem, that includes high model development cost to generate computer-based solutions for energy saving and the cost associated with the implementation of its real-time solutions. The second challenge is the need for quick and rigorous quantification of the benefits from energy saving that can be used to justify the cost of tackling the first challenge. This paper addresses the second challenge through the introduction of new user friendly tool/software for real-time energy targeting and the application of this tool in the rigorous calculation of energy savings that can be realized in a real industrial facility through the application of real-time operations optimization. This user-friendly software for real-time energy integration is used to investigate the potential benefits of energy savings that can be obtained from implementing real-time operation optimization in a typical polycarbonate plant. An academic case study of a plant with two hot streams and two cold streams adapted from El-Halwagi and co-workers 1996 is used to demonstrate the capability of the software and introduce the new methodology.

The industrial case study introduced in this paper, with some modification for the sake of confidentiality, exhibits a middle size chemical plant with more than fifty hot and cold streams in the heat recovery section. However, it demonstrates that huge energy benefits can be obtained upon applying real-time operations optimization for energy saving in polycarbonate processes. It shows that for typical polycarbonate plant operational modifications for the sake of energy saving can result in up to 43% savings in utilities. It also shows that simple control structure of few operating conditions can result in 20% saving in the hot utility. These results would open the door for the wide spread application of more aggressive energy recovery solutions in chemical plants.

TARGETING FOR ENERGY SAVING IN POLYCARBONATE PLANTS USING REAL-TIME ENERGY INTEGRATION

Mahmoud Bahy Noureldin
A.N.M. Kamrul Hasan
Department of Materials and Process Engineering
The University of Waikato,
Hamilton, New Zealand

ACKNOWLEDGMENTS

- The University of Waikato _Emerging Staff Research Award
- Foundation of Research Science and Technology (FRST fund for energy solutions)
- Computer Science UOW

OUTLINE

- Background
    - Mass &Heat integration
- Interval Constraint Logic Propagation
- Problem Statement
- Solution Strategy
- Case Study
- Typical Polycarbonate Plant Study
- Conclusions

BACKGROUND

Process Integration is a holistic approach to process design,retrofitting and operation which emphasizes the unity of the process. It addresses:

- Energy saving
- Pollution prevention
- Reduction of fresh resources(water/raw material usage)
- Reduction of treatment cost(water or air)
- Recovery of valuable chemicals
- Debottlenecking
- Process Simplification
- Enhancement of yield

THERMAL PINCH DIAGRAM

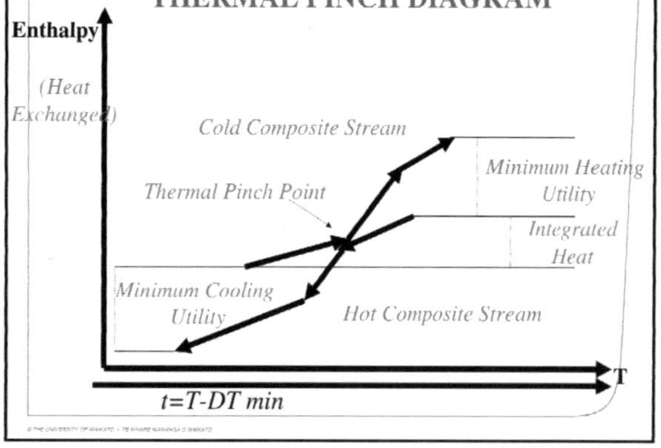

SYSTEMATIC APPROACH

- ❑ **It is dangerous to overlook interactions when using empirical over design values.**
- ❑ **The concept of "worst" operating condition and its selection is far from trivial and what looks like logical can lead to extremely poor design.**
- ❑ **Process variability is a result of infinite combinations of process variations**

- **Optimum process operability can not be reached through adding more units and/or an ad hoc excessive over sizing of process units but rather by systematic design techniques based on a thorough understanding of the system and its physico-chemical and mathematical characteristics.**

PROBLEM STATEMENT

"Given any industrial complex that consists of several processing facilities, determine the best case and worst case scenarios of heating and cooling Utilities under all possible combinations of process modifications and find real-time solution (s) that minimize energy consumption "

$$A.B = [\min\{a_1b_1, a_1b_2, a_2b_1, a_2b_2\}, \max\{a_1b_1, a_1b_2, a_2b_1, a_2b_2\}]$$

For Continuous Unary Operations, $r(x)$

Such as, e^x, $\ln x$, $\sin x$, x^n,..etc

$$r(x) = [\min_{x\in X} r(x), \max_{x\in X} r(x)]$$

CONSTRAINT LOGIC PROPAGATION

Suppose we have such model

$x+y=z$, $y \le x$

and

$x\in[1,10]$, $y\in[3,8]$,
$z\in[2,7]$

Since $x \ge 1$ and $y \ge 3$, CON1 gives $z \ge 4$, so reset the bounds of z to **[4,7]**

Since $z \le 7$ and $y \ge 3$, CON1 gives $x \le 4$, so reset the bounds of x to **[1,4]**

.
.
.

Then
$x\in[3,4]$, $y\in[3,4]$, $z\in[6,7]$

SOLUTION METHODOLOGY

The methodology used in our software to define the global Targets for minimum energy consumptions "under all possible combinations of process modifications" employs a plant energy path model in a constraint logic propagation environment with an interval solver

Academic Case Study

ENERGY TARGETING FOR A PHARMACEUTICAL PLANT

El-Halwagi and co-workers 1996

PLANT DATA

"Given an industrial process with hot streams to be cooled and cold streams to be heated as follows:
H1, with a heat capacity FCph1=10 kW/K,
is to be cooled from 520K to 330K
H2, with FCph2=5 kW/K,
is to be cooled from 380K to 300K
C1, with FCpc1=19 kW/K,
is to be heated from 300K to 550K
C2, with FCpc2=2 kW/k,
is to be heated from 320K to 380K"

OBJECTIVE

Define the minimum consumptions of both the heating and cooling utilities under all possible combinations of process modifications

The ranges of possible combinations of process modifications in the hot and cold streams temperatures and heat capacity flows (FCp) for this pharmaceutical plant are as follows:
H1, with FCph1=(8,15) kW/K
is cooled from (517,522)K to (328,331)K
H2, with FCph2=(3,7) kW/K
is cooled from (378,383)K to (298, 303)K
C1, with FCpc1=(17,20) kW/K
is heated from (297,303)K to (548, 553)K
C2, with FCpc2=(2,3) kW/K
is heated from (318,321)K to (378, 380)K

SOFTWARE RESULTS

Under these process conditions The "absolute minimum target" for heating utility calculated using the CLP-based software is 900 kW, and the "absolute minimum target" for the cooling utility is 12 kW.

The global maximum target for heating utility is 3,605 kW, and the absolute Maximum target for the cooling utility is 126 kW These are the "global" targets attainable with heat integration.

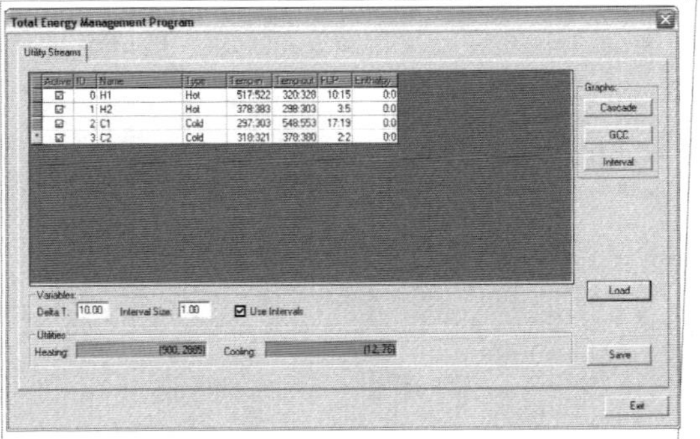

Attainable Best Case Scenario for Heating Utilities

The global minimum heating utility of 900 kW and minimum cooling utilities of 57 kW are realized at the following process conditions:

H1, with a FCph1=15 kW/K, is cooled from 522K to 328K

H2, with a FCph2=7 kW/K, is cooled from 383K to 303K

C1, with a FCpc1=17 kW/K is heated 301K to 548K

C2, with a FCpc2=2 kW/K, is heated from 321K to 378K

It is important to note here that there is another set of process conditions that render the same best case scenario for minimum heating utility of 900 kW:

H1, with FCph1=15 kW/K, is cooled from 522K to 328K

H2, with FCph2=6.7 kW/K, is cooled from 383K to 303K

C1, with FCpc1=17 kW/K, is heated from 303K to 548K

C2, with FCpc2=2 kW/K, is heated from 321K to 378K

Interestingly, the

CLP-based software introduced in this paper indicates that the minimum heating utility target can be realized from two different process conditions.

The later solution cannot be found using the exhaustive enumeration in the current automated pinch techniques of most current commercial software, because these process conditions do not lie on the vertices of the specific heat and temperature flow intervals.

Attainable Best Case Scenario for Cooling Utilities

The global minimum cooling utility consumption of 12 kW and the minimum heating utilities of 1243 kW are realized at the following process conditions:

H1, with FCph1=15 kW/K, is cooled from 522K to 328K

H2, with FCph2=3 kW/K, is cooled from 383K to 303K

C1, with FCpc1=17 kW/K, is heated from 297K to 548K

C2, with FCpc2=2 kW/K, is heated from 321K to 378K

Worst Case Scenario Process energy consumption benchmarks

The "rational" global maximum target for heating utility/ theoretical benchmark is 3,605 kW, and the absolute maximum target for the cooling utility/theoretical benchmark is 126 kW. These are the targets attainable with complete heat integration. However, these two rational targets cannot occur at the same time in this pharmaceutical plant because of the physical nature of the process and its non-linearity. Global maxima for heating and cooling utilities, due to process energy model's non-linearity, are not always simultaneously attainable.

The rational maxima still occur even when intentionally poor process actions are attenuated. In other words, rational maxima represent the maximum energy consumptions for the process due to all possible combinations of process conditions, including disturbance and uncertainty, even while using state-of-the-art heat integration.

Attainable Worst Case Scenario for Heating Utilities

The absolute maximum heating utility of 3605 kW calculated using this novel software is realized at the following process conditions with only 27 kW cooling utility:

H1, with a FCph1=8 kW/K, is cooled from 517K to 331K

H2, with a FCph2=3 kW/K, is cooled from 378K to 298K

C1, with a FCpc1=20 kW/K, is heated 297K to 553K

C2, with a FCpc2=3 kW/K, is heated from 318K to 380K

Attainable Worst Case Scenario for Cooling Utilities

The global/absolute maximum cooling utility consumption of 126 kW is realized at the following process conditions with only 1064 kW heating utility:

H1, with FCph1=15 kW/K, is cooled from 517K to 328K

H2, with Fcph2=7 kW/K, is cooled from 383K to 298K

C1, with FCpc1=17 kW/K, is heated from 303K to 553K

C2, with FCpc2=2 kW/K, is heated from 321K to 380K

Reaction Area:

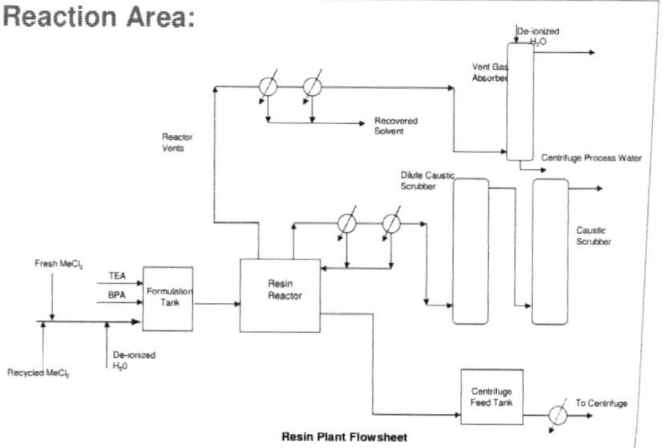

Resin Plant Flowsheet

Centrifuge, Concentrator, and Column Dryer Areas:

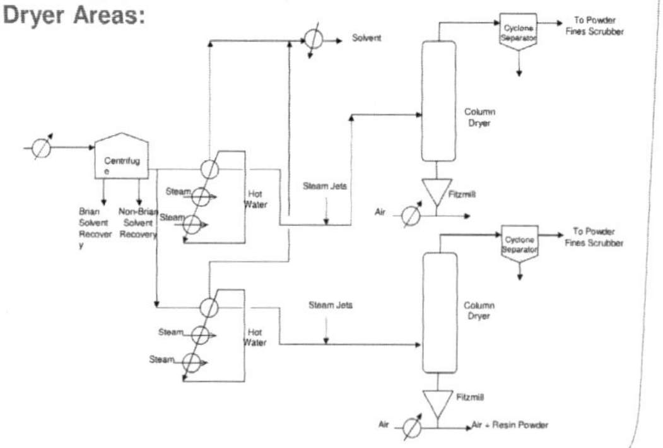

Resin Plant Flowsheet

Powder Fines Scrubber Process:

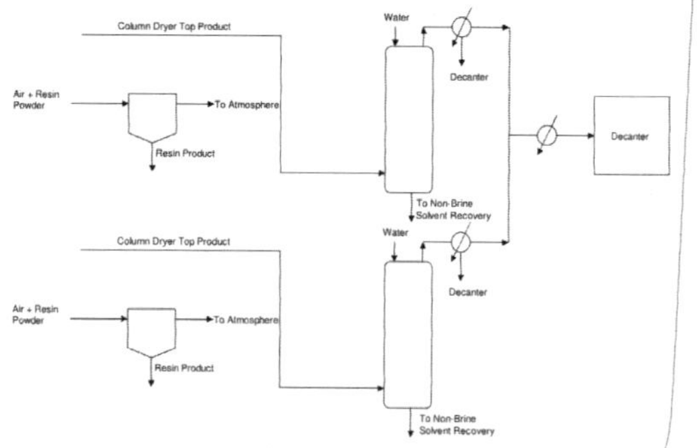

Resin Plant Flowsheet

Post Dryer:

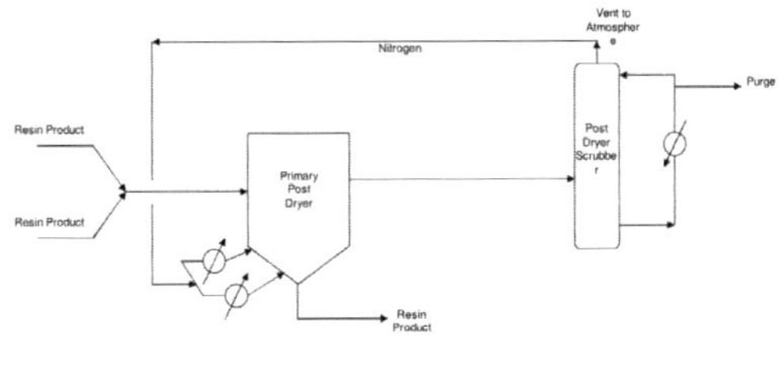

Resin Plant Flowsheet

Solvent Recovery:

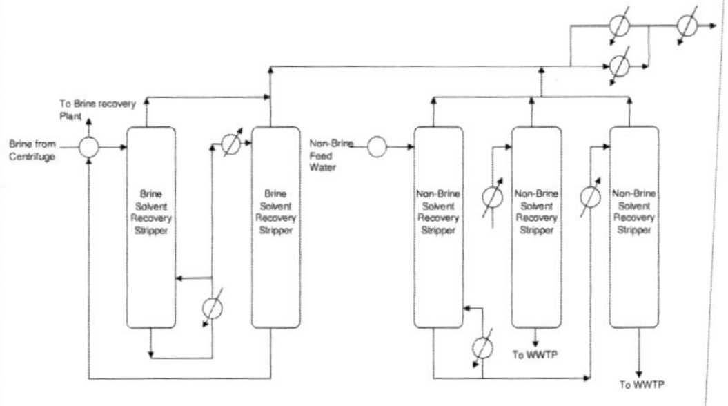

Resin Plant Flowsheet

TEA Recovery:

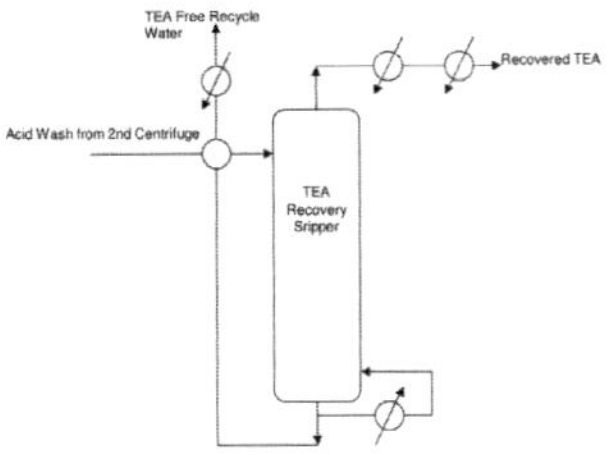

Resin Plant Flowsheet

Minimum Utility Calculated with the Nominal Data:

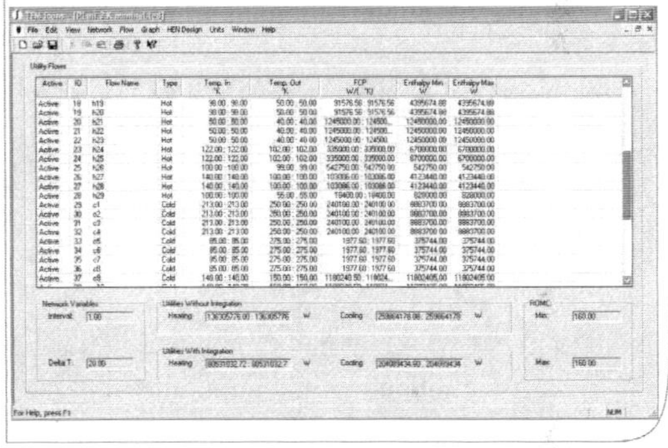

Q_{hmin} with few selective changes:

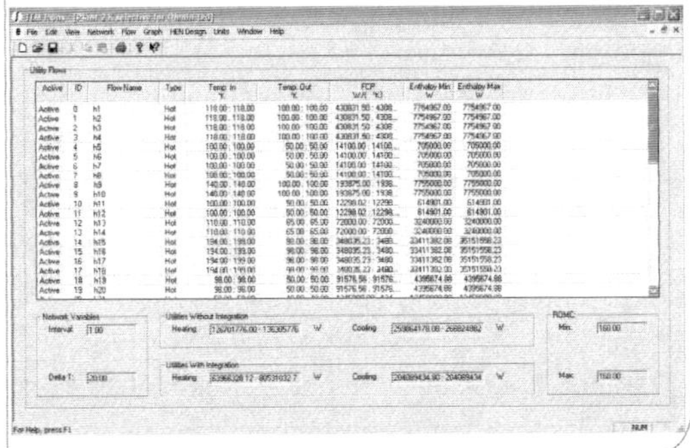

RESULTS DISCUSSION:

- Minimum heating utility requirements, Q_{hmin} 80,531,032 Btu/hr, calculated by Pinch Analysis with ΔT_{min}=20^{0}F using nominal design data
- Heating utility down to 55,984,623 Btu/hr applying ±5^{0}F change from the nominal supply and target temp and savings of Q_{hmin}=43.8%
- Only increasing supply temp of streams H_{15}-H_{18},C_1-C_4, and decreasing the target temp of C_1-C_4, 5^{0}F cause Q_{hmin} to 63,966,328 Btu/hr and contribute 20.56% savings of Heating utility

CONCLUSION

- A CLP-based software is introduced in this paper
- It can be utilized to determine the best case and worst case scenarios of energy consumption under all possible combinations of process plant modifications
- The essence of this new software technology resides in the simultaneous merger of constraint logic propagation, interval mathematics, and process integration principles.
- This novel software technology incorporates and applies a unique targeting tool to establish rigorous targets for energy consumption that can be efficiently utilized in energy resources planning and scheduling

Process Integration for Simultaneous Water and Energy Minimization

Boondarik Leewongtanawit and Jin-Kuk Kim
Department of Process Integration, UMIST, Manchester, UK
PO Box 88, Manchester, M60 1QD, UK

Prepared for Presentation at the 2004 Spring National Meeting, April 2004, "Using Process Integration Technology in Industry to Address Environmental Issues"
Unpublished

Abstract

Water and energy are key commodities utilized in the process industries. Water minimization and energy minimization have been studied separately. A new research direction has drawn attention to a combined analysis of water and energy systems. In this paper, a systematic methodology for simultaneous water and energy minimization is presented to design water and heat exchanger networks to achieve water and energy targets. Whereas water targets are minimum freshwater consumption and wastewater generation, energy targets involve minimum utility consumption, number of heat exchanger units, and heat exchanger area. To implement such a design, two new design aspects are introduced: "Non-isothermal Mixing Point Identification" and "New Separate System Generation".

The first aspect involves "non-isothermal mixing", which enables direct heat recovery between water streams, and therefore allows the reduction of the number of heat transfer units. An MILP model is formulated to identify feasible non-isothermal mixing points, which satisfy minimum freshwater and utility requirements and create an overall water network with fewer number of heat exchanger units. The other aspect is the generation of "separate system" in heat exchanger network design. The flexibility of mixing and splitting of water streams allows separate systems to be created as a series of heat exchanger units between freshwater and wastewater streams. The previous concepts of separate system generation (Savulescu, 1999) have been extended to achieve less number of separate systems and higher temperature driving forces. Consequently, a heat exchanger network of the new method could be designed with fewer heat transfer units and smaller heat transfer area. The new design aspects have been illustrated with an example. Compared with the previous method, the new approach provides a better design with lower utility consumption, fewer heat exchanger units, and smaller heat exchanger area.

1. Introduction

Water is one of the most widely used raw materials in chemical and petroleum industries. Significant amounts of water are required in washing, stripping, and manufacturing processes. As water resources face scarcities, ever-increasing prices, and more stringent environmental regulations, much attention has been paid to reduce freshwater consumption and wastewater generation. There are traditionally two approaches to design these systems: conceptual approach and automated approach. The former analysis exploits graphical tools to explore the possibilities of water reuse (Wang and Smith, 1994), whilst the latter employs mathematical optimization models to obtain a cost-effective solution (Alva-Argaez, Kokossis and Smith, 1999). The analysis of water management generally involves water distribution among water-using operations with the criteria of contaminant concentration levels.

For particular operations such as washing and sterilization operations, temperature of water becomes as important as the quality of water. The water system is now subject to not only the constraints of contaminant concentration levels, but also those of the temperature levels. Water streams need to be heated up or cooled down to satisfy the temperature requirements of the operations, and energy consumption becomes necessary for these heating and cooling tasks. Under these circumstances, both energy and water needs to be considered simultaneously.

Therefore, the problem has become a combined analysis of water and energy systems. The general features of the problem involve a set of water-using operations with specifications of flowrates, temperature and contaminant concentration levels, a selection of water sources with different qualities, and a number of heat transfer units (Figure 1). It is desired to determine water and energy targets and specify the distribution of water among the water-using operations as well as the allocation of heat exchangers between these water streams in order to complete the overall network configuration.

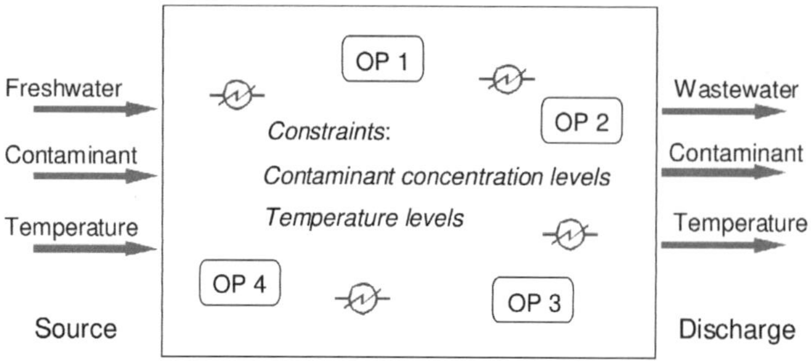

Figure 1. Simultaneous water and energy minimization

In order to obtain an efficient overall design for the problem of simultaneous water and energy minimization, a new methodology has been developed. The method aims to identify a water network with the minimum freshwater and energy requirements and to design a heat exchanger network with minimum number of heat exchanger units and heat transfer area.

2. Review of Simultaneous Energy and Water Minimization

The simultaneous energy and water minimization was first addressed by Savulescu (1999). An example proposed by Savulescu (1999) will be used in this paper to illustrate the presented approach. The data of the example are given in Table 1. The problem is solved under the following assumptions.

(i) Single contaminant operations.
(ii) Single freshwater source (C = 0 ppm and T = 20°C) and single wastewater collection (T = 30°C)
(iii) A minimum temperature approach, ΔT_{MIN}, is imposed to be 10 °C
(iv) Each water-using operation has a fixed temperature and runs isothermally.
(v) The water flowrate through an operation does not change.
(vi) Only reuse options are considered to minimize freshwater consumption.
(vii) Only water streams are considered in the analysis of the heat exchanger network. In other words, non-water using operations are neglected. An equal specific heat capacity for all streams is assumed to be 4.2 kJ/kg°C.

Table 1. The operating data of Example

Operation	Limiting Water Flowrate (kg/s)	Contaminant Mass Load (g/s)	C_{IN}^{MAX} (ppm)	C_{OUT}^{MAX} (ppm)	Temperature (°C)
Operation 1	20	2	0	100	40
Operation 2	100	5	50	100	100
Operation 3	40	30	50	800	75
Operation 4	10	4	400	800	50

The water and energy targets can be set before design. Minimum freshwater usage and wastewater generation can be obtained by applying a conceptual design technique (Wang and Smith, 1994), equal to 90 kg/s of freshwater and 90 kg/s of wastewater. Minimum utility requirement can also be predicted as 3780 kW of hot utility and 0 kW of cold utility.

Savulescu (1999) introduced a two-stage approach as follows:

Stage 1. Two Dimensional Grid Diagram for designing a water network

To design a water distribution network, the author introduced a graphical design tool, Two Dimensional Grid Diagram, which incorporates both temperature and contaminant concentration scales. Then water distribution among the operations is determined from the concept of water mains and certain design rules. These rules are 1) starting the reuse water from the hottest source, 2) connecting with the nearest operations in terms of temperature, and 3) introducing non-isothermal mixing.

Stage 2. Separate System Approach for designing a heat exchanger network

To obtain a simplified heat exchanger network for the water system, the author proposed Separate System Approach. The procedure is 1) constructing the energy composite curves to identify the utility requirements, 2) isothermally mixing water streams on each composite curve to reduce the number of streams within the temperature intervals, and 3) creating maximum number of separate systems between the hot and cold composite curves by fixing the hot composite curve and modifying the cold composite curve to reach maximum degradation of driving force.

Figure 1 represents the solution proposed by Savulescu (1999). The design requires only 6 heat exchanger units. However, the actual utility requirement of the design (hot utility 4265 kW and cold utility 485 kW) is higher than the utility target (hot utility 3780 kW and cold utility 0 kW). This design with small number of heat exchangers could be obtained but with utility penalty.

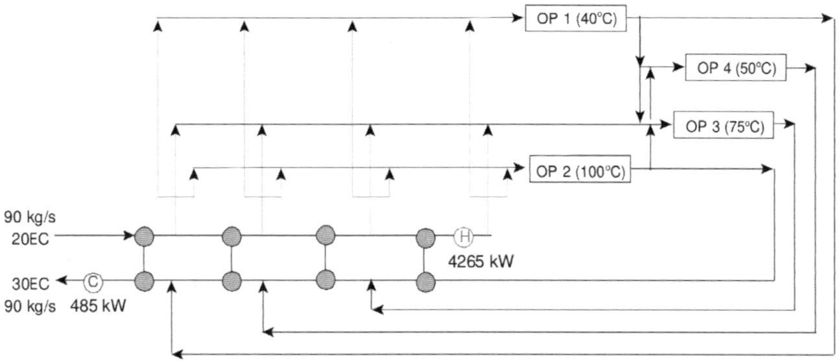

Figure 1. Solution of Example from Savulescu (1999)

This method is a sequential approach that follows a set of design rules in the first stage to provide a water network with less heat exchanger units required. These rules, however, do not always guarantee minimum utility requirement. In the second stage, the author introduced the idea of generating separate systems to simplify a heat exchanger network design. Nevertheless, the generation of separate systems has not been fully explored from the recognition that a smaller number of heat exchanger units could be acquired.

3. New Design Aspects

In order to overcome the aforementioned limitations, a new methodology has been developed to construct a water structure without the utility penalty and provide a better heat exchanger network. The method comprises two new design aspects: Non-isothermal Mixing Point Identification for designing a water network and New Separate System Generation for designing a heat exchanger network.

3.1 Non-isothermal Mixing Point Identification

Non-isothermal mixing is one of possible reuse options in water systems. Instead of indirect heat transfer in a heat exchanger unit (heating a lower temperature stream to be reused in a higher temperature operation or vice versa), direct heat transfer by mixing two streams with different temperatures doesn't need an exchanger (Figure 2). Therefore, the direct heat transfer through non-isothermal mixing can reduce the number of heat exchangers required in the heat exchanger network of the water system.

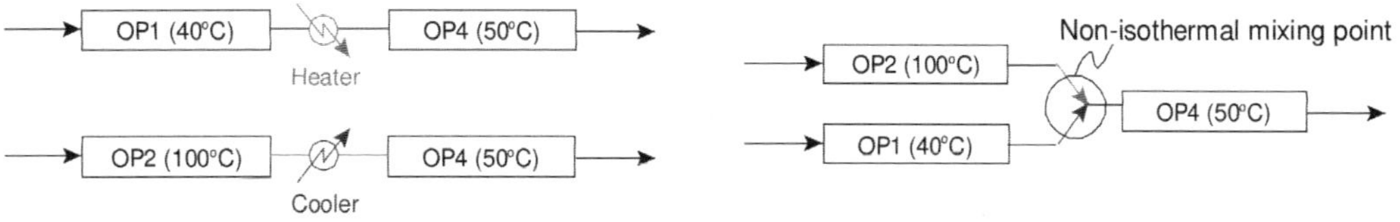

(a) Indirect heat transfer (b) Direct heat transfer

Figure 2. Indirect heat transfer and direct heat transfer options

However, non-isothermal mixing can cause the degradation of temperature driving forces, and also reduces the number of possibilities of indirect heat transfer matching between hot and cold streams. Thus, the arbitrary introduction of non-isothermal mixing may lead to a water network with higher utility requirements as illustrated in the previous section.

The introduction of non-isothermal mixing in the grid design of Savulescu (1999) considers only temperature and contaminant concentration levels, but does not take account of the energy implication as a whole system. Therefore, the resulting water network may require utility consumption different from the utility target. Weighing its pros and cons, non-isothermal mixing in a water network structure needs to be selected carefully.

To identify non-isothermal mixing points that not only satisfy the inlet requirements (temperature and concentration levels) of the operations but also achieve the freshwater and utility targets, a mathematical programming is needed to do the task. It is noted that freshwater and utility targets can be determined before design, as stated in the previous section.

An MILP formulation is based on a combination of modified water superstructure (Alva-Argaez, Kokossis and Smith, 1999) and heat transshipment (Papoulias and Grossmann, 1983) models. Whereas the water superstructure model functions to identify the water network featuring minimum freshwater usage, the heat transshipment model determines indirect and direct heat transfer matches featuring minimum utility consumption. The objective function is to minimize the number of indirect heat transfer matches.

Applying the mathematical programming to the previous example, a new water network, which can achieve both minimum freshwater (90 kg/s) and utility (hot utility 3780 kW) consumption, is identified in Figure 3. The network includes two non-isothermal mixing points (direct heat transfer). One is the mixing of a freshwater stream and two reuse streams at the inlet of Operation 3. The other is the mixing of two reuse streams sent to the same Operation 4. These mixings can reduce the number of heat exchanger units required in the design with out non-isothermal mixing.

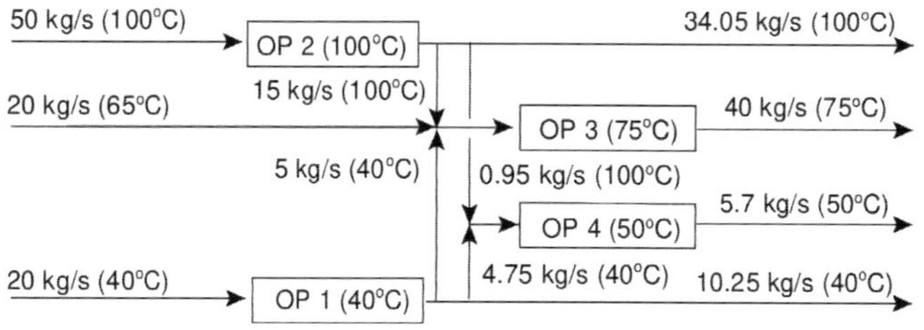

Figure 3. A new water network

After the connections between operations are established, heat exchanger network design is considered to complete the overall network configuration. In the next section, a new separate system approach will be introduced to design the heat exchanger network.

3.2 New Separate System Generation

The heat exchanger network of the water system has distinct features over that of any other fluid system. The heat exchanger network here involved only water streams. The splitting and mixing in the network is allowed not only in the same stream but also between streams, unlike a classical heat exchanger network where the mixing and splitting of different streams are limited. In addition to direct heat transfer via non-isothermal mixing in the previous section, creating separate systems between freshwater and wastewater streams is another way to lower the number of heat exchanger units.

The previous separate system approach (Savulescu, 1999) suggests maximizing the number of separate systems. The separate systems are generated following the shape of the hot composite. Therefore, the number of separate systems is dependent on the number of kink points on hot composite curve, and not necessarily a minimum number of heat exchanger units. Moreover, the cold composite curve is shifted horizontally towards the hot composite curve to maximum degradation of driving force (10°C), and consequently every separate system has minimum temperature driving force. The resulting heat exchanger network of the approach requires unnecessary high number and large area of heat exchangers.

To design a better heat exchanger network for the water system, new separate system generation has been developed. As each separate system represents a heat transfer unit between hot and cold streams, the number of separate systems should be minimized in order to achieve the minimum number of heat exchanger units. Besides, the temperature driving forces should be maximized to reduce heat transfer area, as the more the temperature driving force, the less the heat transfer area. Therefore, the concept of new separate system approach intends to create minimum number of separate systems and maximum temperature driving force in each separate system.

Table 2. Thermal stream data from the water network of Figure 3

Stream	Inlet Temp. (°C)	Outlet Temp. (°C)	Heat Flow Capacity (kW/°C)	Enthalpy (kW)
FW to OP1	20	40	84	1680
FW to OP2	20	100	210	16800
FW to OP3	20	65	84	3780
WW from OP1	40	30	43	-430
WW from OP2	100	30	143	-10010
WW from OP3	75	30	168	-7560
WW from OP4	50	30	24	-480

To illustrate the new separate system approach, the thermal data of streams referred to the new water network (Figure 3) is presented in Table 2. The procedure is based on the following steps:

Step 1. Construct the energy composite curves

The initial energy composite curves are generated based on individual thermal stream data and a minimum temperature approach (10°C) as represented in Figure 4 (a). These composite curves assure that the energy requirements for the new water network achieve the utility target of 3780 kW hot utility and 0 kW cold utility.

Step 2. Minimize the number of separate systems

In order to obtain the minimum number of separate systems, separate systems should be created following kink points on the cold composite curve, as the cold composite curve has fewer number of kink points than the hot composite curve. Then, the boundaries of separate systems are defined at kink points from the selected curve as shown in Figure 4 (b).

Step 3. Maximize temperature driving force in each separate system

The flexibility of the modification of hot and cold composite curves allows minimum number of separate systems to be created with maximum temperature driving forces. Through non-isothermal mixing of hot wastewater streams, the hot composite curve can be modified to maintain maximum driving force in each separate system. Figure 4 (c) illustrates the generation of the separate systems.

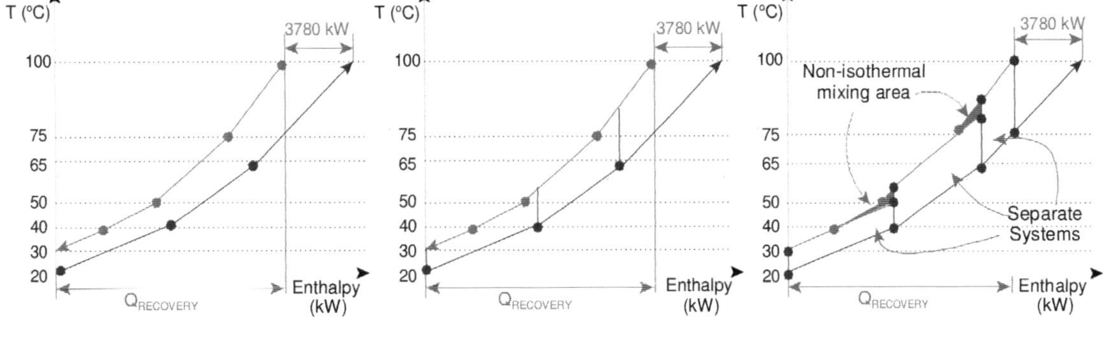

(a) Initial energy composite curves (b) Boundaries of separate systems (c) New separate systems

Figure 4. New separate system approach: Step 1 - 3

Heat loads exchanged between wastewater and freshwater streams in the separate systems are vertically transferred, and the shaded area between the original and the modified hot composite curves represents the non-isothermal mixing of hot wastewater streams.

Step 4. Determine water distribution between separate systems and operations

Since some modifications have been made to the composite curves, water distribution between the separate systems and the operations must be determined. The water distribution involving non-isothermal mixing of wastewater streams is carried out by solving a series of mass and heat balance equations, as presented in Figure 5.

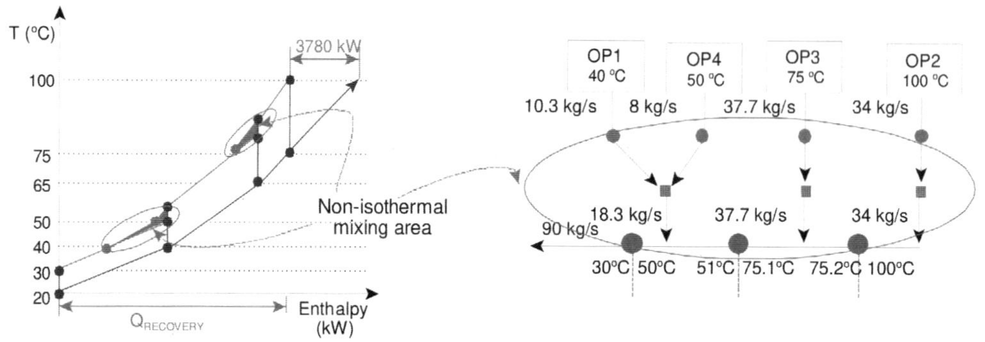

Figure 5. A water distribution between separate systems and operations

The total number of heat transfer units is four, as there are three heat exchangers (separate systems) plus one heater. The new separate system generation provides lower number of separate systems and larger temperature driving forces. As the result, simpler heat exchanger network with fewer heat transfer units and smaller heat transfer area, can be achieved. The new overall network configuration is presented in Figure 6.

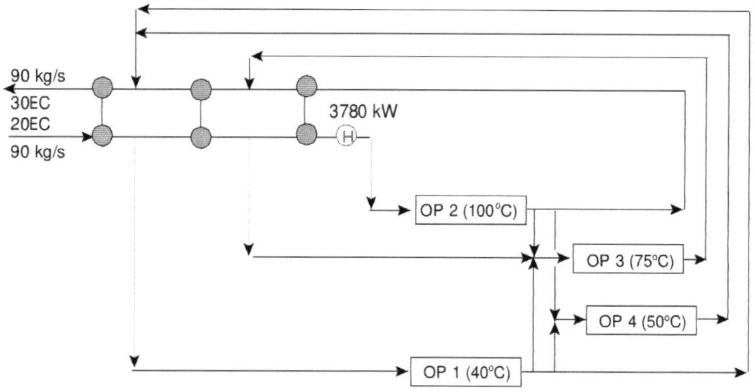

Figure 6. New overall network configuration

A comparison of designs from the previous and new approaches is made in Table 3. The new approach provides a better design with less utility usage, fewer heat transfer units and smaller heat transfer.

Table 3. The result comparison

Requirements	Previous design	New design	%
Freshwater (kg/s)	90	90	-
Hot utility (kW)	4265	3780	11
Cold utility (kW)	485	0	100
Heat exchanger unit (unit)	6	4	33
Heat exchanger area (m^2)	1145	952	17

4. Conclusions

A new approach of the simultaneous water and energy minimization has been presented. The method relies on two sequential design aspects to achieve the water and energy targets. Non-isothermal Mixing Identification is introduced to design an energy-efficient water network. An MILP model is proposed to identify feasible non-isothermal mixing points, which create an overall water network with minimum freshwater and utility consumption. Then, New Separate System Generation is developed to design a simplified heat exchanger network. The new approach provides a heat exchanger network with fewer heat transfer units and smaller heat transfer area.

References

Alva-Algaez A., Kokossis A. and Smith R., 1998, An integrated design approach for wastewater minimization: theory and applications, IChemE Research Event, UK.

Papoulias, S. A. and Grossmann, I. E., 1983, A structural optimisation approach in process synthesis 2: Heat recovery networks, Computers & Chemical Engineer, 19, 71-76

Savulescu L., 1999, Simultaneous Energy and Water Minimization, PhD. Thesis, Department of Process Integration, UMST, Manchester, U.K.

Wang Y. P. and Smith R., 1994, Wastewater Minimization, Chemical Engineering Science, 49, 981-1006.

Automated Design for Efficient Use of Energy in Water Systems

Boondarik Leewongtanawit and Jin-Kuk Kim
Department of Process Integration, UMIST, Manchester, UK
PO Box 88, Manchester, M60 1QD, UK

Prepared for Presentation at the 2004 Spring National Meeting, April 2004, "Using Process Integration Technology in Industry to Address Environmental Issues"
Unpublished

Abstract

A recent research direction has focused on the problems involving both water and energy systems. To take account of the interactions between these systems, a new automated design methodology has been developed. An MINLP model relies on a combined superstructure between water superstructure and heat exchanger network superstructure. The formulation also incorporates two distinct structural features "non-isothermal mixing" and "separate systems" into the superstructure. The objective of the method is to design an overall network structure featuring minimum total annualized cost involved freshwater cost, energy utility cost, and heat exchanger cost. The optimization strategy has been proposed. A simplified MILP model is generated in order to provide an initial point for the MINLP problem. Then, the MINLP model is solved to provide an optimum design of water network and heat exchanger network. This new automated approach has been illustrated with examples, and proven to efficiently provide cost-effective designs as well as practically deal with large size problem.

1. Introduction

Energy and water are vital resources of process industries. Substantial energy is utilized for heating and cooling purposes, while large amounts of water are consumed in various operation tasks. These resources encounter powerful economic and environmental driving forces to rationalize their usages. Therefore, significant efforts have been made to minimize energy and water consumption.

Savulescu (1999) addressed the importance of simultaneous energy and water minimization, and introduced a conceptual approach, which combines the principles of water pinch and energy pinch analysis. The approach is typically a two-stage procedure. In the first stage, a graphical representation is proposed to explore water reuse options and design a network of water distribution among operations. In the second stage, the concept of separate systems is exploited to simplify a heat exchanger network design.

Such a conceptual design method may reach its limitations when dealing with a large number of water-using operations within industrial processes. Also, a wide range of constraints need to be considered in water network designs in practice, such as minimum or maximum flowrates, forbidden matches because of long pipe-runs, and compulsory or forbidden matches for operability. On the other hand, heat exchanger synthesis itself is also a complicated problem. Various aspects need to be considered such as efficient heat recovery, matching, and process practicalities. Therefore, a mathematical programming becomes necessary.

Alternatively, Bagajewicz, Rodera and Savelski (2002) proposed an optimization approach for providing energy-efficient solutions to the water allocation problem in refineries and process plants. The method relies on two sequential LP problems to determine minimum water usage and minimum utility consumption target values. Once these targets are identified, an MILP model is generated to obtain a water network and a corresponding heat exchanger network.

The existing methodologies are sequential approaches of targeting water and energy usages, and then designing water and heat exchanger network structures corresponding to the targets. These methods aim to sequentially minimize freshwater consumption, utility consumption, and number of heat exchanger units. However, the trade-offs between water consumption, utility requirement, number and area of heat transfer units have not been fully considered in the previous approaches.

This paper presents an automated design methodology for the combined water and energy synthesis, which can take account of the simultaneous trade-offs between water and energy systems. The optimization framework is formulated to provide an overall cost-effective design of water network and heat exchanger network.

2. Problem Descriptions

Given is a set of water-using operations, which require water of certain qualities and temperature levels. It is desired to determine a network of water stream interconnections among the operations and to design a network of heat exchangers between these streams. Several assumptions have been made in the study.

In terms of operating data, all water-using operations are single contaminant operations specified by the maximum inlet and outlet concentrations, contaminant mass to be transferred, and operating temperatures. Each water-using operation is assumed to run isothermally with no heat losses or heat gains, and has fixed water flowrate without flowrate losses or flowrate gains.

Another assumption is a single water source available. This water source has been considered to be freshwater with a contaminant concentration level at 0 ppm and a fixed supply temperature of 20°C. No discharge constraint has been made for the effluent. However, the discharge temperature has been specified to be 30°C.

In this system, only reuse options are considered to minimize freshwater consumption. And only water streams are considered in the analysis of heat exchanger networks. This implies an equal specific heat capacity for all the streams (assumed to be 4.2 kJ/kg°C), because the contaminant concentrations are of the order of parts per million.

Table 1 shows the data of an example introduced by Savulescu (1999), which will be used to illustrate the presented approach.

Table 1. Example 1 from Savulescu (1999)

Operation	Contaminant Mass Load (g/s)	C_{IN}^{MAX} (ppm)	C_{OUT}^{MAX} (ppm)	Temperature (°C)
Operation 1	2	0	100	40
Operation 2	5	50	100	100
Operation 3	30	50	800	75
Operation 4	4	400	800	50

3. System Interactions

In case of water network synthesis, the task is to identify water consumption, wastewater generation, and water network configuration for a particular set of water-using operations. The problem is subject to the constraints of contaminant mass load and concentration levels, where the typical objective function is a minimum freshwater consumption. There may be more than one water structure with the same water usage. For Example 1, two water networks presented in Figure 1 require the same 90 kg/s of freshwater.

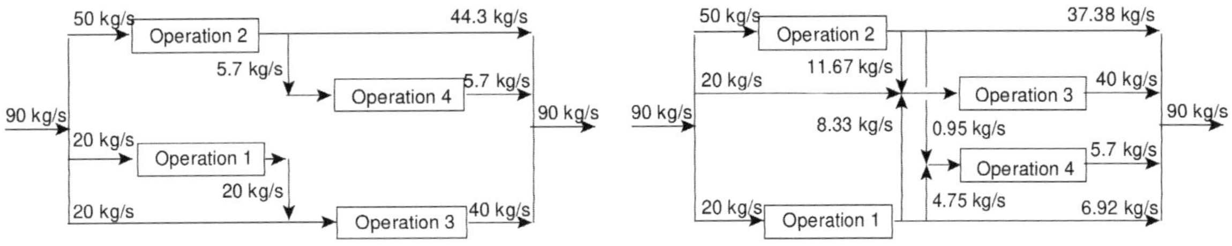

(a) Water network I (b) Water network II

Figure 1. Two water networks with minimum freshwater consumption

When energy aspects are considered in the water networks, not only those contaminant restrictions need to be regarded, but also the temperature constraints of operating conditions. Water streams require energy consumption for heating or cooling to satisfy the temperature requirements of the operations. Figure 2 shows the energy implications of two different water structures.

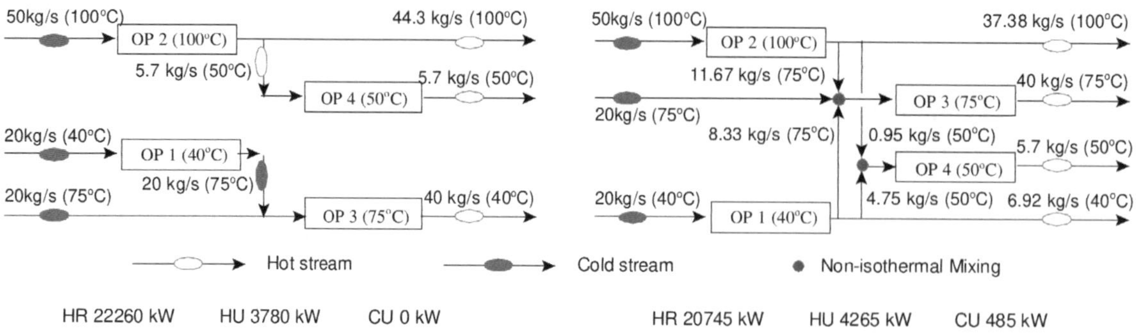

(a) Water network I

(b) Water network II

Figure 2. Energy implications in water networks

Different water network configurations can achieve the target for minimum water consumption, but each structure involves different energy implications. Some influences of water structure on heat integration are important to point out as follows:

- *Sets of hot and cold streams:* For the traditional heat exchanger network synthesis, sets of hot and cold streams and their heat capacity flowrates are clearly defined from the start. In the case of a heat exchanger network in water systems, the heat capacity of streams, however, relies on the flowrate of water stream connections. Besides, the temperature conditions of sinks and sources of each water stream indicate whether the stream is hot or cold stream.

- *Heat recovery (HR):* A variety of reuse options result in dissimilar heat transfer schemes (i.e. direct or indirect heat recovery). Indirect heat recovery between hot and cold streams in a heat exchanger unit is one of heat recovery options. In the other hand, direct heat recovery through mixing of streams with different temperatures is also possible. The potential of heat recovery is a function of water reuse alternatives.

- *Utility consumption (HU and CU):* The utility requirement is directly dependent to the flowrate of water through the overall system as well as the water supply and discharge temperatures. The more water usage, the more utility consumption. Besides, in case of the same water usage, utility consumption can vary according to the structure of water network.

There are clearly interactions between water and energy systems and the heat exchanger network heavily depends on the design of water network selected. When the energy aspects are not taken into account during the construction of water network, the overall network design may result in inefficient use of energy and capital cost.

4. An Automated Approach

In order to systematically consider energy aspects in the design of water systems, an automated approached is developed. The objective function is to minimize a total annualized cost, which is the summation of 1) freshwater cost, 2) utility cost, and 3) heat exchanger cost. The mathematical model is based on a combined superstructure of water network and heat exchanger network. The model formulation involves three main elements: water superstructure, heat transshipment model and heat exchanger network superstructure.

Water superstructure (Alva-Argaez, Kokossis and Smith, 1999) involves freshwater consumption and wastewater generation together with the connections of water streams, whilst heat transshipment model (Papoulias and Grossmann, 1983) identifies utility requirements and heat transfer matches between hot and cold streams. Finally, heat exchanger network superstructure (Floudas, Ciric and Grossmann, 1986) determines network configurations including stream interconnections, stream flowrates and temperatures and areas of heat exchanger units.

Figure 3 shows a simple representation of the combined superstructure of water and heat exchanger networks with two water-using operations. In the superstructure, two distinct structural features are noted in this study.

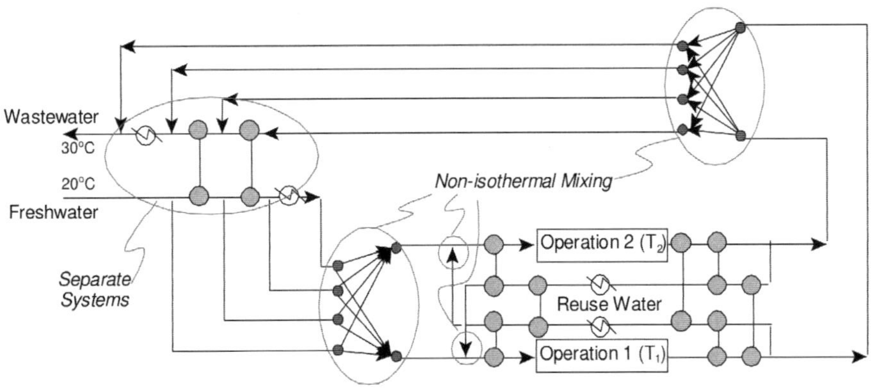

Figure 3. A representation of the combined superstructure

1) *Non-isothermal mixing:* In water systems, heat loads between hot and cold streams can be recovered indirectly or/and directly. Direct heat transfer via non-isothermal mixing offers another degree of freedom to the design of heat exchanger networks. Despite the degradation of temperature driving forces, non-isothermal mixing can reduce the number of heat transfer units required in the heat exchanger network, as it can replace indirect heat transfer in an exchanger. Therefore, non-isothermal mixing has a potential to reduce the capital cost. However, both direct heat transfer and indirect heat transfer are included in the superstructure as heat recovery alternatives, which will be selected in a cost-effective manner during the optimization.

2) *Separate Systems:* The flexibility of mixing and splitting of water streams allows the design of heat exchanger network to be simplified. As freshwater streams come from the same source of water, freshwater can be progressively heated up and split to feed different operations. On the other hand, since wastewater streams are sent to the same discharge, they can be mixed and cooled down sequentially in the opposite way. These mixing and splitting allow separate systems to be created as a series of heat exchangers between cold freshwater and hot wastewater streams. Therefore, simple configuration of heat exchanger network can be obtained with less number of heat exchangers units. The concept of separate systems has been extended and implemented into the optimization framework, which will automatically identify the number of separate systems, the location of mixing and splitting, and temperature conditions of heat exchangers.

As a result of the combination, the mathematical model is formulated as an MINLP problem. Due to the nonlinearity, a simplified MILP is first solved to provide a good initial point. The MILP model is obtained from setting outlet concentrations to their maximum values and the relaxation of heat exchanger cost. Then, the MINLP model is optimized by using the solution of the MILP as the initial point and provides a cost-effective design of water and heat exchanger networks

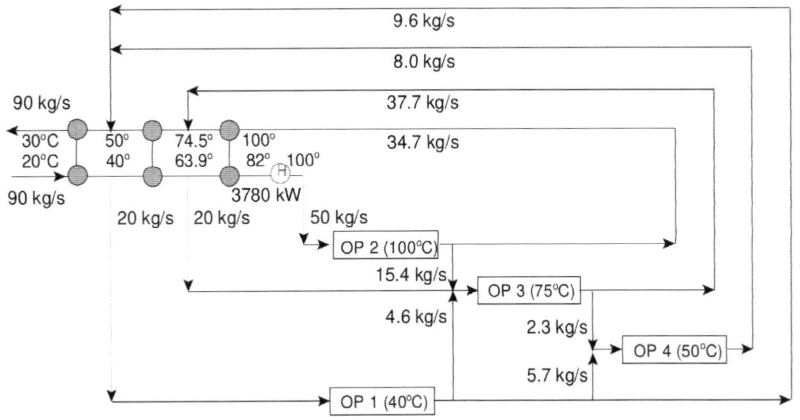

Figure 4. The minimum total cost design for Example 1

Applying the optimization technique to Example 1, a minimum total cost design of water and heat exchanger networks is illustrated in Figure 4. Compared with the conceptual approach proposed by Savulescu (1999), the automated approach can offer an optimal design with less utility usage, fewer heat transfer units and smaller heat transfer area and makes a 12% saving of the total annualized cost. The comparison of the designs for Example 1 is shown in Table 2.

Table 2. The result comparison of Example 1

Requirements	Conceptual design	Automated design
Freshwater (kg/s)	90	90
Hot utility (kW)	4265	3780
Cold utility (kW)	485	0
Heat exchanger unit (unit)	6	4
Heat exchanger area (m^2)	1145	939
Total cost (£/yr)	972,000	860,000

5. Case Study

A case study proposed by Bagajewicz, Rodera and Savelski (2002) includes 8 water-using operations. The data of Case Study is presented in Table 3.

Table 3. Case Study from Bagajewicz, Rodera and Savelski (2002)

Operation	Contaminant Mass Load (g/s)	C_{IN}^{MAX} (ppm)	C_{OUT}^{MAX} (ppm)	Temperature (°C)
Operation 1	2	25	80	40
Operation 2	2.88	25	90	100
Operation 3	4	25	200	80
Operation 4	3	50	100	60
Operation 5	30	50	800	50
Operation 6	5	400	800	90
Operation 7	2	400	600	70
Operation 8	1	0	100	50

The presented approach can also be applied with this large size problem. As a result, an overall network configuration featuring minimum total annualized cost for Case Study is illustrated in Figure 5.

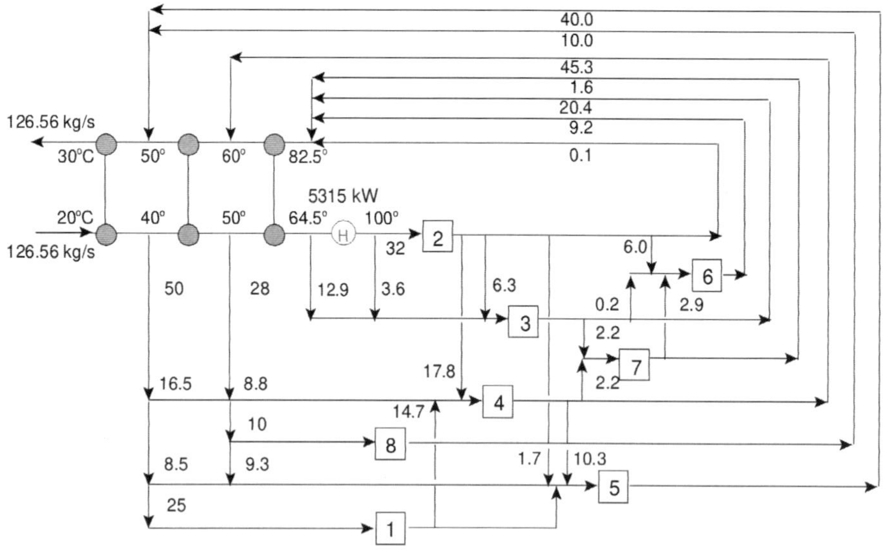

Figure 5. The minimum total cost design for Case Study

Table 4 shows a comparison between the previous design (Bagajewicz, Rodera and Savelski, 2002) and the new design by the developed automated approach. The new design provides 7% saving of total annualised cost when compared with the previous design. It should be noted although the new design requires slight increases of freshwater and utility consumption; it contributes a considerable reduction of heat exchanger units. This demonstrates that the simultaneous trade-offs between freshwater and utility consumption, and network configuration have been performed through the proposed method.

Table 4. The result comparison of Case Study

Requirements	Previous design	New design
Freshwater (kg/s)	125.94	126.56
Hot utility (kW)	5290	5315
Cold utility (kW)	0	0
Heat exchanger unit (unit)	12	4
Heat exchanger area (m^2)	1011	899
Total cost (£/yr)	1,230,000	1,160,000

6. Conclusions

An automated design methodology for the combined water and energy synthesis has been presented. The approach exploits a mathematical programming based on a combined superstructure between water superstructure and heat exchanger network superstructure. The formulation also accommodates non-isothermal mixing and separate system generation in the structural features. The objective function is to design an overall network structure of water network and heat exchanger network featuring minimum total annualized cost. The automated approach has proven to effectively deal with complex problems with large number of operations.

References

Alva-Algaez A., Kokossis A. and Smith R., 1998, An integrated design approach for wastewater minimization: theory and applications. IChemE Research Event, UK.

Bagajewicz M., Rodera H. and Savelski M., 2002, Energy efficient water utilization systems in process plants. Computers and Chemical Engineering, 26, 59-79.

Floudas C. A., Ciric A. R. and Grossmann I.E., 1986, Automatic synthesis of optimum heat exchanger network configurations. AIChe Journal, 32, 276-290.

Papoulias S. A. and Grossmann, I. E., 1983, A structural optimisation approach in process synthesis 2: Heat recovery networks, Computers & Chemical Engineer, 19, 71-76.

Savulescu L., 1999, Simultaneous Energy and Water Minimization. PhD. Thesis, Department of Process Integration, UMST, Manchester, U.K.

Wang Y. P. and Smith R., 1994, Wastewater Minimization. Chemical Engineering Science, 49, 981-1006.

Optimal Abnormal Situations Management in Chemical Processes through Mass Integration

*Noureldin, M.B., Swan, J.E., and **Shelley, M.D.
*Materials and Process Engineering Department,
University of Waikato, Hamilton, New Zealand
** University of Houston, TX, USA

Abstract:

Mass integration as a holistic approach to the generation, depletion, separation, allocation and routing of species and streams in process plants has come of age. It began in the late 1980s as a new systematic technique for targeting and synthesizing mass exchange networks in environmental applications. Since then mass integration has evolved and branched into a powerful technology for industrial applications. As a systematic methodology for process plant synthesis, design, operation and optimization, mass integration facilitates the fundamental understanding of the global mass flow within any process and exploits this understanding to identify performance targets. Optimization of the allocation, separation, and generation of streams and species is utilized to achieve these desired performance targets. El-Halwagi and Spriggs (CEP 1998) provide a thorough background of general mass integration principles. Process utilities, such as cooling water, steam, electricity, fuel, nitrogen, hydrogen, air, caustic soda, and waste water systems, are at the core of any industrial chemical process. Furthermore, disturbance diversion from the main process paths to the utility paths represents the industrial state-of-the-art to control process variability and uncertainty against undesired production quality. Due to the constraints imposed by environmental regulations in recent decades, water treatment facilities have also become a critical component of process utility systems, which must be continuous and robust. For example, a sudden failure in a process utility such as fresh water flow or water bio-treatment capacity could result in plant shutdowns and production losses. In this paper, a systematic methodology to optimally manage these situations using mass integration techniques is introduced. The first step in applying mass integration methodologies to abnormal utility situations is to develop a global mass allocation model of the whole process from a species viewpoint. A simple, rationally generated process superstructure, which includes mass integration strategies such as stream segregation, mixing and recycling, can provide additional design and operational degrees of freedom that can be exploited during abnormal utility situations. In addition, each process sink/generator can be manipulated via changes in design and/or operating parameters, such as flow rates and compositions, to avoid process shutdowns due to abnormal utility situations.

This model can be augmented within an interval constraint satisfaction model to first target for the achievable production and quality targets subject to different abnormal situation scenarios of a given utility system. A back calculation approach is then adapted using given production and quality targets to obtain different solutions that can optimally respond to abnormal utility situations. The methodology and model development will be illustrated with theoretical and industrial case studies. The academic case study introduced is an adapted chemical plant from El-Halwagi and co-workers, 1996. The simple model used to illustrate the methodology is taken from pollution prevention through process integration book of El-Halwagi, 1996.

Optimal Abnormal Situations Management in Chemical Processes through Mass Integration

M. Bahy Noureldin, Janis Swan
Materials and Process Engineering Department,University of Waikato, Hamilton, New Zealand
Mark Shelley
University of Houston, TX, USA

Outline

- Background and Motivations
- Problem Statement
- Problem Formulation
- Tools
- Technology
- Solution Strategy
- Case Study(s)
- Conclusion

BACKGROUND

Mass Integration is a holistic approach to process design, retrofitting and operation which emphasizes the unity of the process. It addresses:

- Pollution prevention
- Reduction of fresh resources (water/raw material usage)
- Reduction of treatment cost (water or air)
- Recovery of valuable chemicals
- De-bottlenecking
- Process Simplification
- Enhancement of yield
- Optimal process operation management

Problem Statement

"Given an existing process, identify
Its attainable targets and optimal solutions
that meet a set of process objectives.
The devised process strategies shall
incorporate integrated design and
operational modifications.
The targets and the optimal solution
scenarios
should address specified abnormal situations
due to disturbances and uncertainties"

Objectives

- Simultaneous Targeting for different process objectives such as, Maximum Quality; Minimum Pollution;Minimum Utilities;Maximum yield;...etc.
- Develop Optimal Design and Operational Strategies to Achieve these Targets under different specified levels of abnormal situations due to disturbances and uncertainties

Solution Strategy

_Calculation of the attainable
targets of the process objectives under all possible
combinations of design and operational modifications
While considering the effect of
disturbances and uncertainties using
Constraint Logic Propagation with Interval
Labels Techniques

_Searching for the optimal design
and operating solution that achieve the desired target(s)
using process integration techniques and/or
mathematical programming

Synthesis of Optimal Operating Policies Under Disturbances and Uncertainties

El-Halwagi and co-workers 1996

Ethylene Chloride Plant

The University of Waikato

Ethyl Chloride Plant

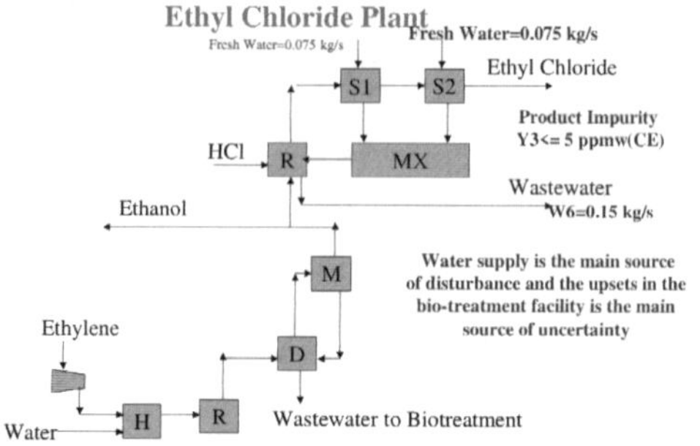

The University of Waikato

SOLUTION STRATEGIES

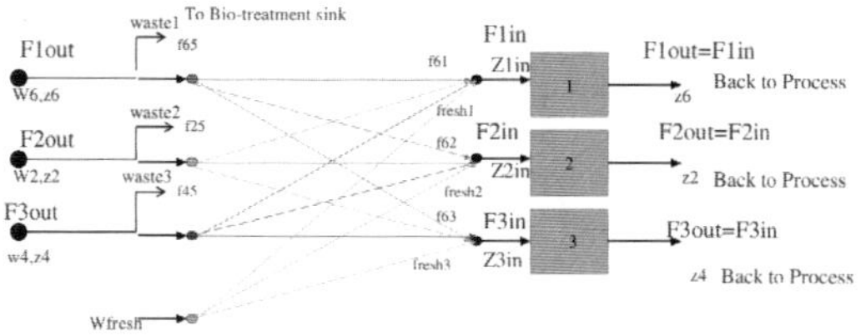

The University of Waikato

Mathematical Formulation

Pollution Load=(f65*z6+f25*z2+f45*z4)

Total Utilities=fresh1+fresh2+fesh3

Overall Mass Balance equations(Mixers)

F1in-f21-f41-f61-Fresh1=0.0;

F2in-f22-f42-f62-Fresh2=0.0;

F3in-f23-f43-f63-Fresh3=0.0;

Overall Mass Balance Equations(Splitters)

F1out-f65-f61-f62-f63=0.0;

F2out-f25-f21-f22-f23=0.0;

F3out-f45-f41-f42-f43=0.0;

The University of Waikato

Component Mass Balance Equations(Splitters)

F1in*Z1in-f21*z2-f41*z4-f61*z6=0.0;

F2in*Z2in-f22*z2-f42*z4-f62*z6=0.0;

F3in*Z3in-f23*z2-f43*z4-f63*z6=0.0;

Units Overall Material balance

F1in-F1out=0.0;

F2in-F2out=0.0;

F3in-F3out=0.0;

The University of Waikato

Unit Operations Performance Model Equations

(0.15+0.2*F1in)*y1-(F1in-0.09)*Z1in=6.03;

y1-5*z6=0.0;

0.15*(y1-y2)-F2in*(z2-Z2in)=0.0;

(F2in/0.015)^1.3-(1-0.015/F2in)*(y1-0.1*Z2in)/(y2-0.1*Z2in)-0.015/F2in=0.0;

(F3in/0.015)^1.3-(1-0.015/F3in)*(y2-0.1*Z3in)/(y3-0.1*Z3in)-0.015/F3in=0.0;

0.15*(y2-y3)-F3in*(z4-Z3in)=0.0;

The University of Waikato

Unit Operations Operating Windows

Z1in<=65.0;

Z1in>=0.0;

Z2in<8.0;

Z2in>=0.0;

Z3in=0.0;

F1in<=0.15;

F1in>=0.09;

F2in<=0.09;

F2in>=0.075;

F3in<=0.085;

F3in<=0.075;

Non-negativity constraints

y2>=0.0;

y1>=0.0;

y3>=0.0 ;

f41>=0.0;

f42>=0.0;

f43>=0.0;

f45>=0.0;

f61>=0.0;

f62>=0.0;

f63>=0.0;

f65>=0.0;

Fresh1>=0.0;

z2>=0.0;

z4>=0.0;

z6>=0.0;

f21>=0.0;

f22>=0.0;

f23>=0.0;

f25>=0.0;

Fresh2>=0.0;

Fresh3>=0.0;

PROCESS ATTAINABLE TARGETS

- **Minimum Pollution target=0.0 kg/s of waste water**
- **Minimum Fresh resources target=0.0 kg/s of fresh water**
- **Maximum Quality target=0.25 PPM of CE as an impurity in the end product**

Maximum Quality Optimal Design Solution

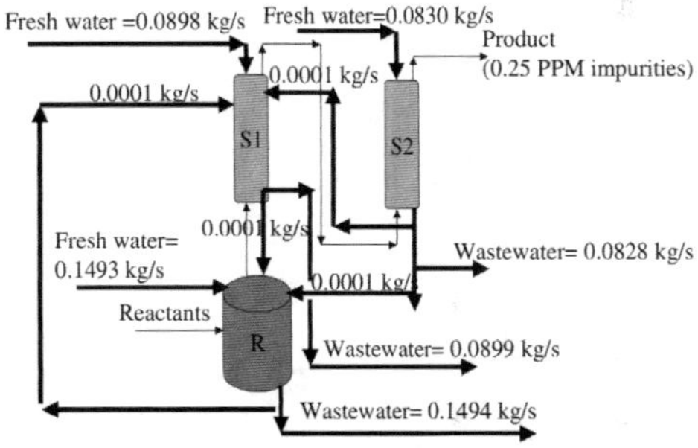

Optimal Operating Policy # 1

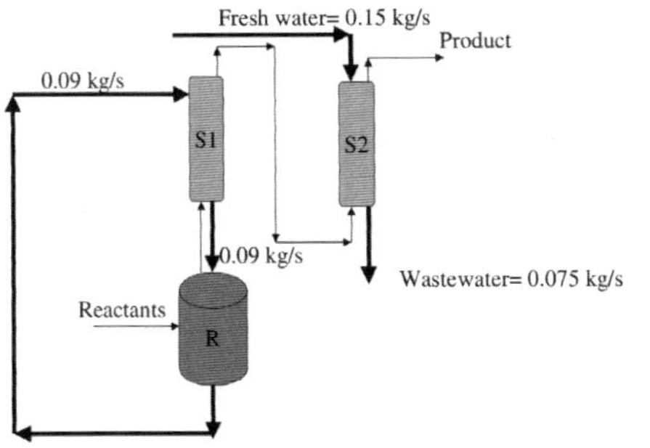

Ethyl Chloride Plant

Optimal Operating Policy # 1

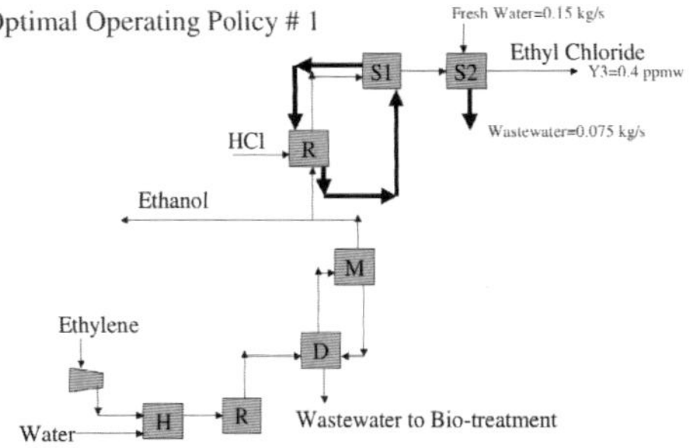

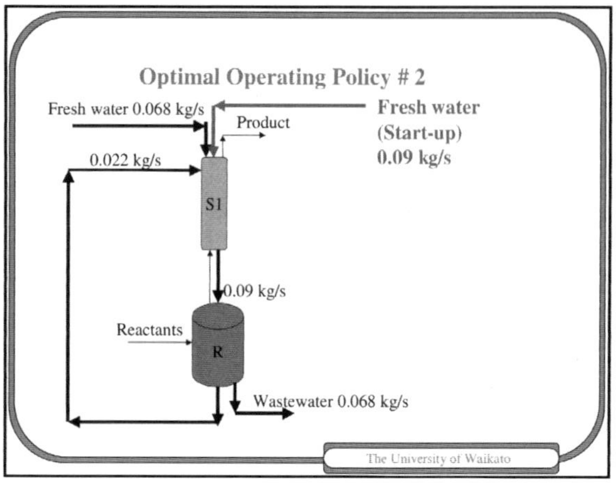

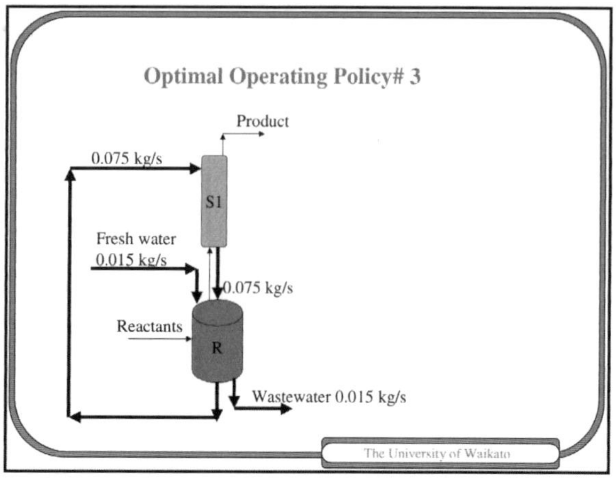

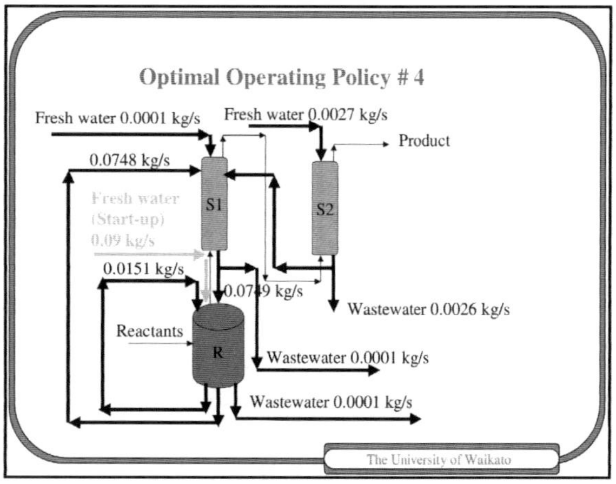

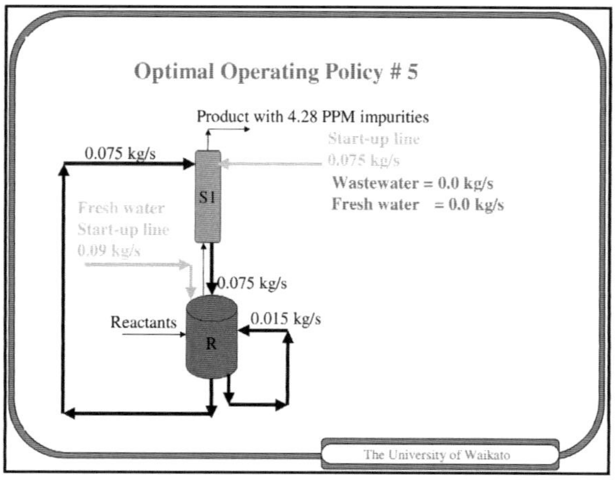

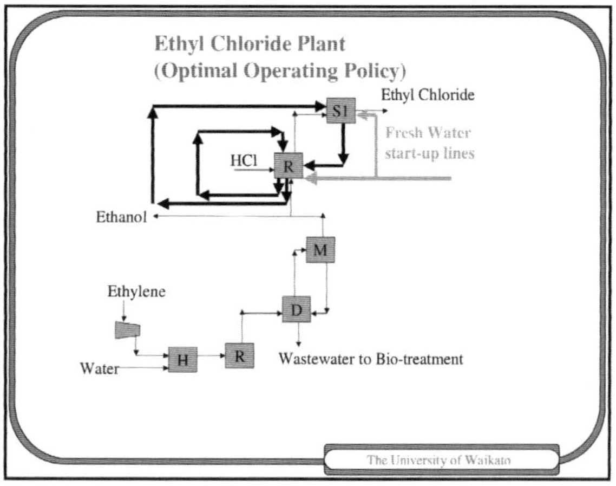

CONCLUSION

- Systematic tool for simultaneous targeting of process objectives
- Optimal design and operating Policies for reconciling process objectives via identifying optimal process solutions to achieve the desired targets under disturbances and uncertainties
- Short-cut method for generating solution strategies for any combination of process targets

The University of Waikato

Implementation of Chemical Recycling, Water Conserving in the Metal Plating Industry in the U.S.-Mexico Border Region.

By

Jacek Dziewinski* ++, Gary Scott**, Eli Maestas***, Johnny Harper*, Ned Elkins*

Paper presented at the Spring Meeting of the American Institute of Chemical Engineers, New Orleans, April 25-29, 2004.

Session: Application of Process Integration Tools for Energy Efficient Design and Water Minimization Design

UNPUBLISHED

* *Los Alamos National Laboratory, P.O. Box 1663, Los Alamos, NM 87545, Mail Stop J587.*
** *Department of Energy, Carlsbad Field Office, P.O. Box 3090, Carlsbad, NM 88221.*
*** *Applied Science Laboratory, P.O. Box 21158, Albuquerque, NM 87154*
++ *Corresponding author*

Abstract

The United States Department of Energy-Carlsbad Field Office through its National Border Technology Partnership Program (NBTPP) is promoting and aiding pollution prevention programs within the US-Mexico border region. Water resources in this region are very scarce on both sides of the border. Metal plating shops are heavy water consumers and they are a very well known source of pollution. They use copious amounts of municipal water to fill their plating baths and their rinsing tanks. The water flows through their operations and then is discharged as wastewater. Wastewater from plating operations contains heavy metals as well as organic compounds and cyanides. There are well over a hundred plating shops in Mexico in the border region. For most, their in-house expertise in chemistry and engineering is limited and they operate based on tradition and operational experience.

A project is under way, which, should all metal platers adopt conservation methods, would result in the conservation of 1,600 acre feet (562 million gallons) of fresh municipal water per annum and eliminate generation of over 450 million gallons of wastewater. Implementing process modifications and introduction of improved process control methods and installation of separation equipment will achieve this. The challenge of introducing such controls lies in the fact that the water conservation and pollution prevention projects in Mexico must be economically driven. Unlike in the US (where these activities are mainly regulation driven, and possible costs savings are mostly in avoidance of high waste treatment costs and of potential fines or closures) in the Mexican border region, the owners must realize a clear economic advantage in recovery and savings of the costs of chemicals they use for their operations, in addition to adhering to Mexican environmental norms and standards.

The project is piloted at a medium size plating shop Electracobados y Servicios Industries in Ciudad Juarez in partnership with NBTPP and Los Alamos National Laboratory. The shop employs about 90 personnel, processes approximately 30,000 lbs. of metal parts per day and it uses and discharges almost 6 million gallons of water per year.

The production process was analyzed, and mathematical models for the concentration profiles were developed. The process analysis showed that the amount of water that flows through the operations is fixed, independently on production volume, or of the number or size of the parts they plate. The results of the modeling were written into an excel spreadsheet, which computes the necessary water consumption depending on production volume. This model demonstrates that by adjusting the water flow according to the production level could reduce water consumption and consequently wastewater generation by up to 75%. Use of this model in combination with simple methods of process control could result in such substantial water conservation and pollution prevention.

However, this approach does not provide an economical incentive to the Mexican factory. Water bills are low and the same amount of chemicals will be lost to waste water, although at higher concentrations. Only real savings in money spent on chemicals will cause the platers to embrace and implement the changes.

The plating operation consists of a plating bath, where chemicals are fed, and a series of rinsing tanks, through which the plated parts are moved in order to rinse excess salts from their surface. The city water flows counter-current to the movement of the processed parts. Consequently, the first rinsing tank, adjacent to the plating bath, has the highest concentration of the chemicals, and the last rinse tank (which receives fresh city water) has the lowest concentration of the impurities. The water from the first rinse tank, containing the highest concentration of the chemicals, is discharged as wastewater resulting in the loss of chemicals it contains.

Concentrating the chemicals in the wastewater flowing out of the first rinse tank, and returning them to the plating bath will result in their recovery, while purified waste water will be reused for rinsing. After evaluating ion exchange, reverse osmosis, and evaporation systems, it has been decided that a spray chamber evaporator would be the most economical solution. Attached to the plating bath it will utilize the heat of electrical resistance of the plating and it will also reduce the required cooling burden. The concentrated bottoms will be returned back to the plating bath, while the condensed water will be reused for rinsing. Such closed loop system will concentrate any impurities that come with chemicals, make up water, or water introduced for equipment flushing or bath replacement. Therefore, a city water cleanup system is required for all fresh water introduced to the plant. This pretreatment system consists of an ion exchange softener, a carbon filter, and a reverse osmosis unit.

The materials balances and flow diagrams for the new green plating operation have been prepared. Mathematical models for the concentration profiles have been developed and built into an excel spreadsheet. The model allows for concentration profiling and equipment sizing for any operation and production throughput. The required equipment selection has been done and the project is now in the procurement stage.

The estimated recovery of the capital investment for the equipment is expected to occur in less that one year. It is anticipated that after a successful demonstration of the chemical recovery system at Ciudad Juarez other plating shops in the border region will obtain loans from environmental aid institutions, such as North American Development Bank to finance their own implementations to be shortly recovered by chemical savings. The environmental impact on the border region promises to be very significant.

Acknowledgement.

The authors would like to acknowledge the sponsorship of this project by the United States Department of Energy-Carlsbad Field Office through its National Border Technology Partnership Program (NBTPP).

UNCLASSIFIED

Implementation of chemical recycling, water conserving in the metal plating industry.

Jacek Dziewinski, Gary Scott, Eli Maestas, Johnny Harper, Ned Elkins

NNSA Los Alamos

UNCLASSIFIED

UNCLASSIFIED

Presentation Contents

- Goal of the DOE National Border Technology Partnership Project
- Process Optimization of Plating Shops
- Electroacobados Industriales
- Optimization @ Electroacobados
- Flow Diagram and Material Balances
- Optimization by Process Control
- Recycling Concepts
- Selection of a Recycling ethod
- Cost and Environmental Benefits
- Summary & Conclusions

NNSA Los Alamos

UNCLASSIFIED

Project Goal

- Introduce chemical recycling, water conservation systems in the US-Mexico border region
- Design and install and startup a chemical recycling water recovery system at the plating shop in Ciudad Juarez
- Evaluate performance and economics
- Showcase the results
- Assist in implementations at other border businesses

Process Optimization of Plating Shops

- Shops are small to medium establishments
- Operated based on tradition and consultation
- Do not have their own engineering, chemical expertise within their companies
- Operate on small profit margin (every $ counts)
- In order to introduce good practices need help
  - Consulting
  - Capital
  - Must be convinced based on examples

ELECTROACOBADOS Y SERVICIOS INDUSTRIALES

A metal plating shop in Ciudad Juarez.

Size of shop

- 30,000 Lbs. Of Metal Parts Processed Per day.
- 90 employees
- 20 m 3 of solid waste generated per month
- 240 m3 / year
- 20 gallons / minute of water
- 1,200 gallons / hour
- 19,200 gallons / day (16hr work day)
- 480,000 gallons / month (25 day work month)
- 5,760,000 gallons / year of waste water dumped into sewage
- There are 115 similar metal finishing shops in the Border Region

PROCESS OPTIMIZATION AT ELECTROACOBADOS Y SERVICIOS INDUSTRIALES

A stepwise approach is taken:

- system will be installed in one of the plating lines and serve as a pilot for an overall plant recycling system
- Such approach results in overall savings, provides the necessary empirical information for the whole scale design, eliminates overdesign of the full system
- reduces the incidence of possible future full-scale failures a designer is likely to encounter
- stepwise approach will not carry excessive costs in the first evaluative stage of the project.

Material Balance – Zn Line

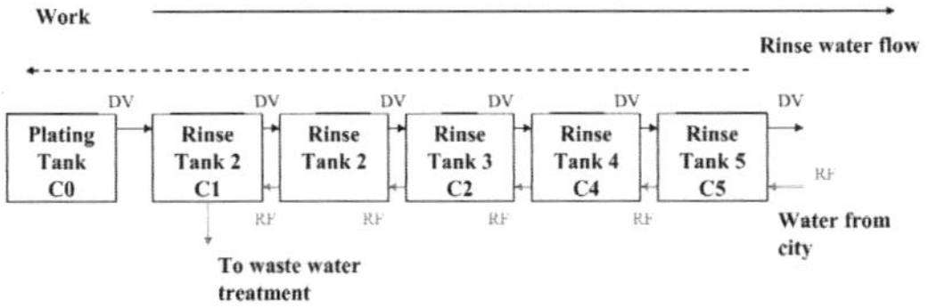

Material Balance for the Zinc Plating Line.

- Rinse Tank #5: DV · C4 = (DV + RF) · C5 (equation 1)
- Rinse Tank #4: DV · C3 + RF · C5 = (RF + CV) · C4 (equation 2)
- Rinse Tank #3: DV · C2 + RF · C4 = (RF + CV) · C3 (equation 3)
- Rinse Tank #2: DV · C1 + RF · C3 = (RF + CV) · C2 (equation 4)
- Rinse Tank #1: DV · C0 + RF · C2 = (RF + CV) · C1 (equation 5)

DV = Drag Volume; RF = Rinse Flow, C = concentration

Balance quations used in excel enable calculation of the concentration profile based on production parameters

$$C5 = \frac{DV \cdot C0}{(DV+RF)\left\{\frac{(DV+RF)^2}{DV^3}\left[\frac{(DV+RF)^2}{DV} - RF\right] - \frac{(DV+RF)^2}{DV^3}RF - \frac{RF}{DV^2}\left[\frac{(RF+DV)^2}{DV} - RF\right]\right\} - RF\left\{\frac{DV+RF}{DV^2}\left[\frac{(DV+RF)^2}{DV} - RF\right] - \frac{DV+RF}{DV^2} \cdots\right.}$$

$$C4 = \frac{DV+RF}{DV} C5$$

$$C3 = \left[\frac{(RF+DV)^2}{DV} - RF\right]\frac{C5}{DV}$$

$$C2 = \left\{\frac{DV+RF}{DV^2}\left[\frac{(DV+RF^2)}{DV} - RF\right] - \frac{DV+RF}{DV^2}RF\right\}C5$$

$$C1 = \left\{\frac{(DV+RF)^2}{DV^3}\left[\frac{(DV+RF)^2}{DV} - RF\right] - \frac{(DV+RF)^2}{DV^3}RF - \frac{RF}{DV^2}\left[\frac{(RF+DV)^2}{DV} - RF\right]\right\}C5$$

INPUT

Rinse Flow (l/min)	8
Barrel/h	8
Drag (l/barrel)	4
Drag Volume	0.5333

Bath concentration	g/l
Zn	37.5
K	64.2
Cl	77.1
NH4	10.1
Surfactant	40
Brightener	0.01
Fe	0.15
Cu	0.1
H2O2	0.125
C	0.3
MnO4	4.2

OUTPUT at Rinse Flow 8 L/min

g/L	Rinse Tank 1	Rinse Tank 2	Rinse Tank 3	Rinse Tank 4	Rinse Tank 5
Zn	2.50	0.17	0.0118	0.000737	4.61E-5
K	4.27	0.28	0.0202	0.001262	7.89E-5
Cl	5.12	0.34	0.0242	0.001516	9.47E-5
NH4	0.67	0.05	0.0032	0.000199	1.24E-5
Surfactant	2.67	0.18	0.0126	0.000786	4.91E-5
Brightener	0.006	4.44E-5	3.15E-6	1.97E-7	1.23E-8
Fe	0.01	0.006	4.72E-5	2.95E-6	1.84E-7
Cu	0.0067	0.0004	3.15E-5	1.96E-6	1.23E-7
H2O2	0.008	0.0055	3.93E-5	2.46E-6	1.53E-7
C	0.02	0.0013	9.44E-5	5.89E-6	3.68E-7
MnO4	0.28	0.018	0.00132	8.259E-5	5.162E-6

INPUT

Rinse Flow (l/min)	2
Barrel/h	8
Drag (l/barrel)	4
Drag Volume	0.5333

Bath concentration	g/l
Zn	37.5
K	64.2
Cl	77.1
NH4	10.1
Surfactant	40
Brightener	0.01
Fe	0.15
Cu	0.1
H2O2	0.125
C	0.3
MnO4	4.2

OUTPUT atRinse Flow 2 L/min

g/L	Rinse Tank 1	Rinse Tank 2	Rinse Tank 3	Rinse Tank 4	Rinse Tank 5
Zn	9.99	2.65	0.84	0.18	0.037
K	17.1	4.54	1.43	0.30	0.063
Cl	20.5	5.45	1.72	0.36	0.076
NH4	2.69	0.71	0.22	0.047	0.010
Surfactant	10.65	0.71	0.89	0.19	0.039
Brightener	0.002	0.0007	0.0002	4.70E-5	9.89E-8
Fe	0.04	0.01	0.003	0.0007	0.0001
Cu	0.026	0.007	0.002	0.0004	9.89E-5
H2O2	0.033	0.0088	0.003	0.0006	0.0001
C	0.08	0.021	0.0067	0.0014	0.0003
MnO4	1.19	0.297	0.0937	0.0197	0.0041

UNCLASSIFIED

Water Conservation by Process Control

- The material balance enables calculation of the concentration profile throughout the plating line at varying work loads
- By proper adjustments of the rinse water flow water consumption may be reduced by 75% without adversely affecting product quality

NNSA — Los Alamos

UNCLASSIFIED

RECYCLING CONCEPTS

Goal:

- Recycle valuable materials (economic incentive)
- Minimize or eliminate waste generation
- Conserve water

This goal can be accomplished by extracting valuable chemicals from waste streams and recycling them back to the process and purifying process water

Means:

- Reverse Osmosisis
- Ion Exchange
- Evaporation
- Combination of the above unit operations

RO - WITH PLATING TANK OVERFLOW - RINSE TANK #3 PRMEATE RETURN

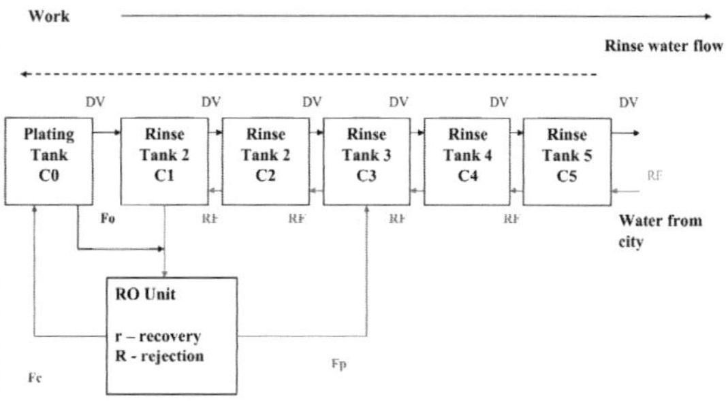

Excel tool calculates the concentration profile E.g. for RO - RINSE TANK 3 PERMEATE RETURN

Input:

Cpt	3750	Metal Concentration in Plating Tank [ppm]
DV	0.5	Estimated Drag Volume [l/min]
R	0.98	Rejection
r	0.75	Recovery
Ff	16	R/O Feed [l/min]

Output:

Permeate Flow [l/min]	Fp	12
Concentrate Flow [l/min]	Fc	4
Plating Tank Overflow [l/min]	Fo	3.5
Makuup City Water Flow [l/min]	Fm	0.5

C1 [ppm]	1680	R/O Feed conc. Cf [ppm]	9515
C2 [ppm]	247	Permeate Conc. Cp [ppm]	190
C3 [ppm]	190	Concentrate conc. [ppm]	37492
C4 [ppm]	127		
C5 [ppm]	63		

C5 as function of R/O Feed – Return to RT#3

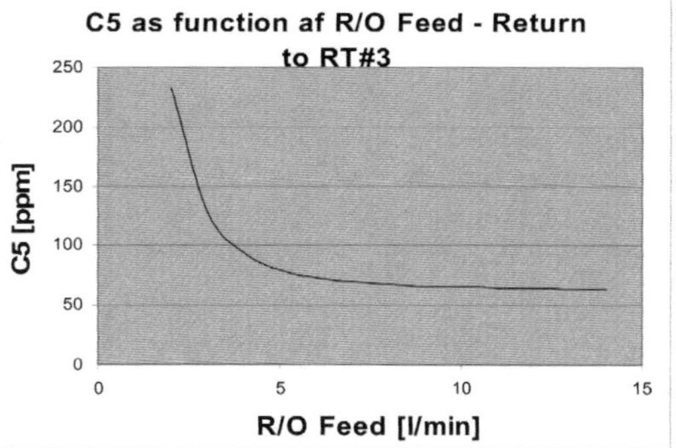

Ion Exchange

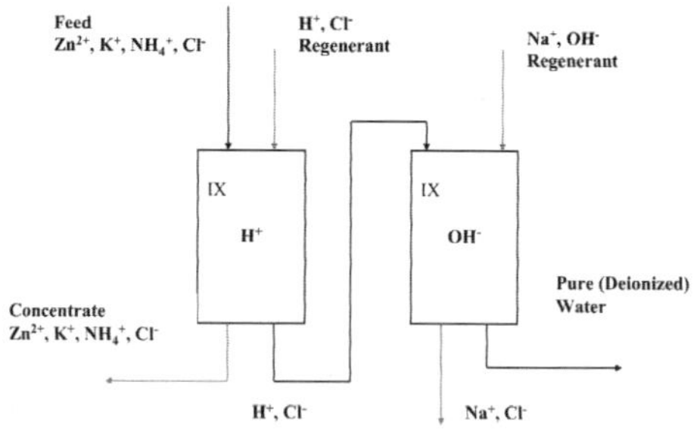

$Zn^{++} + 2Cl^- + 2\,R\text{-}H = R\text{-}Zn\text{-}R + 2H^+ + 2Cl^-$

$H^+ + Cl^- + R\text{-}OH = R\text{-}Cl + OH^- + H^+$

Ion Exchange System

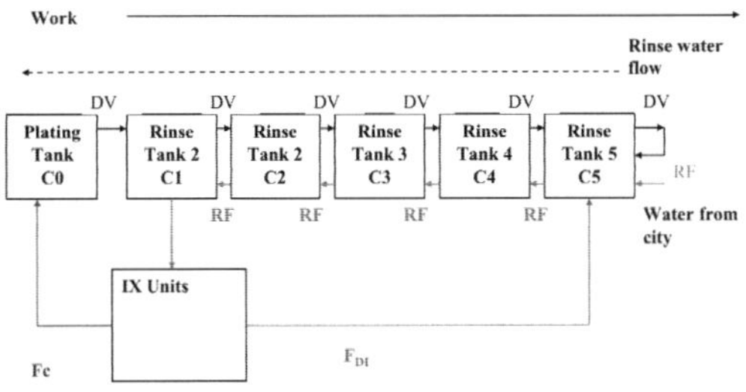

- Does not address organics
- Generates NaCl waste
- Organics adversely affect the resin

Evaporation

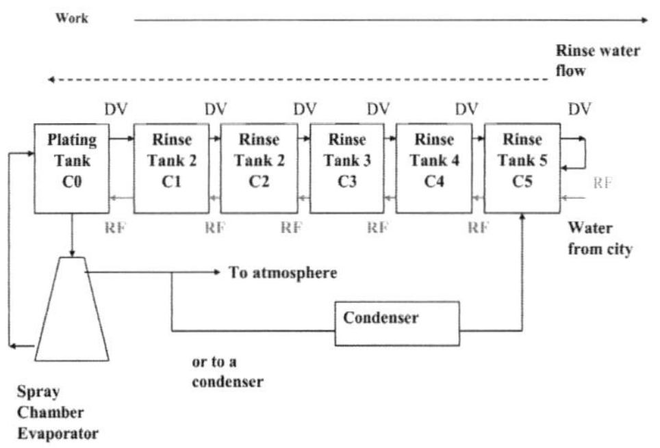

Evaporation system with water pretreatment

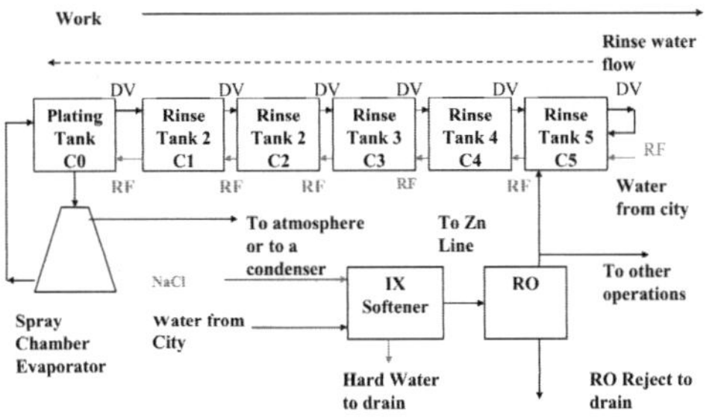

$$CaCO_3 + 2R_1\text{-}Na + 2R_2\text{-}Cl = R_1Ca\text{-}R_1 + R_2\text{-}CO_3\text{-}R_2 + NaCl$$

Cost Benefits

Chemical conservation savings per year:

Zn Barrel line	$29K
Zn Rack line	$29K
Ni line	$36K
Sn line	$18K
Cr line	$18K
Cu line	$34K
Water treatment costs	$21K
Waste sludge disposal costs	$21K
Total costs savings:	$206K

EQUIPMENT COSTS

RO / IX System for make-up water, whole shop	$35K
Evaporator for Zn Barrel line chemical recycling	$7 K
Piping and instrumentation for Zn Barrel line	$8 K
Total equipment cost for pilot implementation at Zn Barrel line	$50K

Total estimated equipment costs for implementation at entire shop (five more lines plus degreasing and acid rinsing): $120K

BENEFITS OTHER THAN DIRECT CHEMNICAL COSTS

- Greatly reduced waste water
- Elimination of solid waste
- Water conservation
- Largely reduced environmental liabilities
- Reduced community health risks
- Elimination of carcinogen discharge into environment
- Conservation of resources
- Good will for our communities
- Better product quality
- Less rework

Cost & Environmental Benefits

	Year 1 1 plant	Year 2 5 plants	Year 3 20 plants	Year 4 40 plants	Year 5 60 plants	Year 6 80 plants	Year 7 100 plants	Year 8 115 plants	Year 9 115 plants	Year 10 115 plants
Capital Investment K$	$170	$680	$2,550	$3,400	$3,400	$3,400	$3,400	$,2250	0	0
Operating costs	$5K	$25	$100K	$200	$300	$400	$500	$575	$575	$575
Chemical Savings	$206	$1,030	$4,120	$8,240	$12,360	$16,480	$20,600	$23,690	$23,690	$23,690
Balance K$	$31	$325	$1,470	$5,640	$8,660	$12,680	$16,700	$20,865	$23,115	$23,115
Water savings Mln gal/year	4.6	23	92	184	276	368	460	529 1,623 acre-ft	529	529
Solid waste avoidance M^3 / year	240	1,200	4,800	9,600	14,400	19,200	24,000	27,600	27,600	27,600

UNCLASSIFIED

Summary

- A recycling technology can be deployed at plating shops in the Border Region at equipment costs paid back by chemicals savings in less than one year
- A piloting effort has been undertaken at Electroacobados y Servicios Industrales in Juarez
- Design tools to be used at each establishment have been developed and are available
- A design package and an economic evaluation package that will require only minor modifications for particular establishments will be available for all plating shops

NNSA Los Alamos

UNCLASSIFIED

A Superior Fe/ZSM-5 Catalyst for Selective Catalytic Reduction (SCR) of Nitric Oxide with Ammonia **

*Gongshin Qi and Ralph T. Yang**

[*] Ralph T. Yang
Department of Chemical Engineering, University of Michigan
Ann Arbor, Michigan 48109, USA
Fax: 1-734-764-7453
E-mail: yang@umich.edu

[**] This work was partially funded by EPRI and NSF.

Introduction

Nitrogen oxides (NO, NO_2 and N_2O) in the exhaust gases from combustion of fossil fuels remain a major source for air pollution. They contribute to photochemical smog, acid rain, ozone depletion, and greenhouse effects. The current technology for reducing nitrogen oxides emissions from power plants is selective catalytic reduction (SCR) of NO_x (x=1, 2) with ammonia in excess oxygen. Many catalysts have been reported to be active for this reaction, such as vanadia, other transition metal oxides (e.g. CuO. Fe_2O_3, Cr_2O_3, Fe_2O_3-Cr_2O_3, etc.), pillared clays and zeolites. [1, 2] The mechanism of the reaction on the vanadia catalysts has been studied extensively and several different mechanisms have been proposed. [3-8] Although the SCR technology based on vanadia catalysts has been commercialized, problems still remain, e.g., high activity for oxidation of SO_2 to SO_3, formation of N_2O at high temperatures and toxicity of vanadia. Hence there are continuing efforts in developing new catalysts.

H-zeolite and ion-exchanged molecular sieves have received much attention for SCR of NO_x by both hydrocarbon and ammonia in recent years. Zeolite based catalysts, such as H-form, Cu-exchanged and Fe exchange zeolites [9-19] were investigated by many groups. Among the catalysts, the Cu-ZSM-5 and Fe-ZSM-5 received much attention because of their high activities, especially for the Fe-ZSM-5 because this catalyst is highly resistant to SO_2 and H_2O under SCR reaction conditions. Several preparation methods for Fe-ZSM-5 were reported. Feng and Hall [11] reported an overexchanged Fe-ZSM-5 prepared by ion-exchange in aqueous solution of oxalate salt. This catalyst was much more active than Cu-ZSM-5 for SCR of NO with isobutene as reductant and also was resistant to sulfur dioxide; but unfortunately, their results were difficult to reproduce.[12] Chen and Sachtler [13] studied the same reaction on Fe-ZSM-5 prepared by sublimation of $FeCl_3$ on H-ZSM-5. They reported high activities and durability for SCR of NO with hydrocarbons. Fe-ZSM-5 prepared by solid-state ion exchanging method with $FeCl_2$,[20, 21] $FeCl_3$[22-24] and $FeSO_4$ [25] as the iron source was also reported. In our previous work, [14, 15] a series of Fe-exchanged molecular sieves prepared by aqueous ion exchange with $FeCl_2$ as the iron source were studied for SCR of NO with ammonia. It was found that Fe-ZSM-5 showed the highest activities among all known catalysts for the reaction. Long and Yang [26] also studied SCR of NO with ammonia over different Fe-ZSM-5 catalysts prepared by different preparation methods. It was found that Fe-ZSM-5 prepared by aqueous ion exchange have higher activities than that prepared by solid state exchange and sublimation of $FeCl_3$. More recently, EPRI [27] reported that the costs for preparation of Fe-ZSM-5 catalysts were still high compared to commercialized vanadia catalyst although Fe-ZSM-5 has high activities and other advantages. In this work, we developed a new and very simple method for preparing Fe/ZSM-5 catalyst by conventional incipient-wetness impregnation method using the NH_4-ZSM-5 and $FeCl_2$ and $Fe(NO_3)_3$ as starting materials. The resulting catalyst prepared from $FeCl_2$ yielded significantly higher activities than catalysts prepared by all other methods. Evidence for ion exchange taking place during the preparation procedure is shown.

Experimental

The volume of the aqueous solution that was used was the same as the total pore volume of the zeolite. Upon contact the solution was spontaneously imbibed by the porous zeolite. The sample was then calcined in air at 500°C for 6 hours. The sample was ready for SCR activity test. By this method, minimum amount of water is evaporated and no excess solution is needed. Also, no washing is required compared to

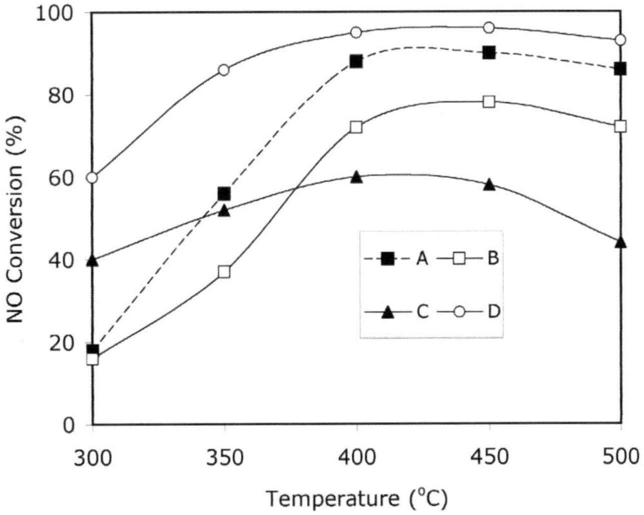

Figure 1. Catalytic activities for NO reduction by ammonia on different catalysts. Reaction conditions: 40 mg catalyst, 1000 ppm NO, 1000 ppm NH_3, 2% O_2, balance He, and GHSV = 5.7×10^5 h^{-1}(ambient conditions) A) Fe-ZSM-5, prepared by ion-exchange in $FeCl_2$ solution; B) 2.5%Fe/ZSM-5, prepared by impregnation with $Fe(NO_3)_3$ as precursor; C) 4.4%V_2O_5-8.2%WO_3/TiO_2 prepared by impregnation; D) 2.5%Fe/ZSM-5, prepared by impregnation with $FeCl_2$ as precursor.

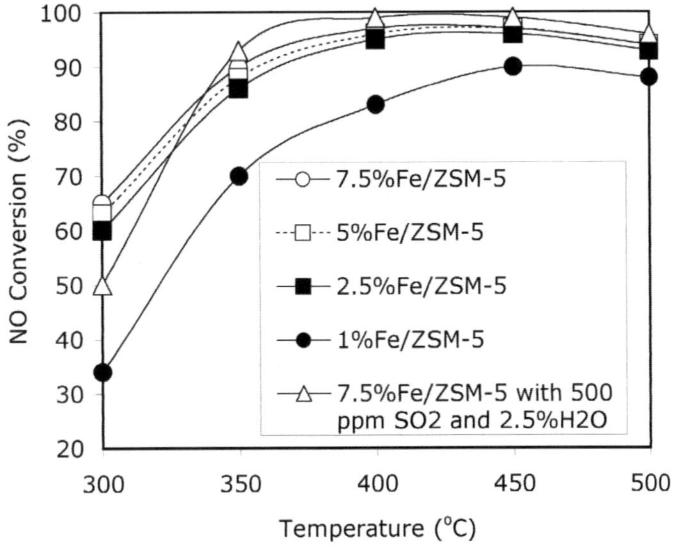

Figure 2. Catalytic activities for NO reduction by ammonia on Fe/ZSM-5 catalysts with different Fe loadings.Reaction conditions: 40 mg catalyst, 1000 ppm NO, 1000 ppm NH_3, 2% O_2, 500 ppm SO_2 (when used), 2.5% H_2O (when used), balance He, and GHSV = 5.7×10^5 h^{-1}.

aqueous ion exchange method. So this method should be less costly than the ion-exchange method using aqueous solution. In order to compare the activities, other Fe-ZSM-5 catalysts and the commercial catalyst 4.4%V_2O_5-8.2%WO_3/TiO_2 were also investigated.[28] The catalyst was also characterized by XRD, ESR and XPS. The powder X-ray diffraction (XRD) measurements were carried out with a Rigaku Rotaflex D/Max-C system with Cu Kα (λ = 0.1543 nm) radiation. The XPS experiment was carried out on a Perkin-Elmer PHI 5400 ESCA system at room temperature under 10^{-8}-10^{-9} Torr, using Mg Kα radiation. ESR spectra was recorded on a Brucker EMX ESR spectrometer, under the conditions of microwave of 0.202 mW and a modulation amplitude of 1.0 G. Before recording the spectra, the samples were purged with He to remove weakly adsorbed oxygen and other impurities. The catalytic activity was tested with a conventional fixed bed flow reactor. The NO and NO_2 concentrations were continually monitored by a chemiluminescent NO/NO_x analyzer (Thermo Environmental Instruments Inc. Model 42C). To avoid errors caused by the oxidation of ammonia in the converter of the NO/NO_x analyzer, an ammonia trap containing phosphoric acid solution was installed before the sample inlet to the chemiluminescent analyzer. The products were also analyzed by a gas chromatograph (Shimadzu, 8A) at 50 °C with 5A molecular sieve column for N_2 and Porapak Q column for N_2O.

Results and Discussion

It is found the catalyst has superior activities for SCR of NO with ammonia in the presence of excess oxygen. The activities are much higher than that on Fe-ZSM-5 prepared by aqueous ion exchange method. The results of NO conversions with different catalysts are shown in Figure 1. Surprisingly high NO conversions (at a very high space velocity) were obtained on the Fe/ZSM-5 catalyst prepared by impregnation with $FeCl_2$ as the iron source. At temperatures over 350 °C, on 2.5%Fe/ZSM-5, the NO conversion reached nearly 90% and also with a wide temperature window similar to the Fe-ZSM-5 prepared by aqueous ion-exchange method. Compared to the Fe-ZSM-5 catalyst prepared by aqueous ion-exchange method, the new Fe/ZSM-5 catalyst has superior activities, especially in the lower temperature range. The catalysts prepared by impregnation with $Fe(NO_3)_3$ as precursor has lower activities, which is consistent with the findings of Long and Yang. [14, 15] In comparison, the commercial catalyst 4.4%V_2O_5-8.2%WO_3/TiO_2 showed substantially lower activities in NO reduction under the same conditions, and the NO conversion dropped more rapidly when the temperature was over 450 °C due to oxidation of ammonia by oxygen. [1, 2] In addition, on the new Fe/ZSM-5 catalyst, no N_2O was detected in the whole temperature range, with only N_2 and H_2O as the products. On the 4.4%V_2O_5-8.2%WO_3/TiO_2 catalyst, about 5% of N_2O was observed in the reaction at 400 °C. NO conversions on Fe/ZSM-5 with different Fe loadings are shown in Figure 2. It was found that the activity increased with increasing Fe loading, but the increase diminished when the Fe loading was > 2.5%. Because the combustion gases usually contain water vapor and SO_2, we further studied the effect of H_2O and SO_2 on the catalytic performance of 7.5%Fe/ZSM-5. As shown in Figure 2, H_2O and SO_2 actually enhanced the SCR activity at temperature above 350 °C. Such enhancement is similar to the Fe-ZSM-5 prepared by aqueous ion exchange method, and has been explained. [14, 15]

Figure 3 shows the XRD patterns of different Fe/ZSM-5 catalysts with different Fe loadings. All the Fe/ZSM-5 catalysts exhibited the typical lines of ZSM-5 zeolite, indicating that the structure of the zeolite remained intact after the treatment. However, the intensity of ZSM-5 observed changes depending on the loading of iron. The decrease of the intensity for the Fe/ZSM-5 catalyst was related to the increasing loading of iron. The decrease of the intensity could be the result of the higher absorption coefficient of iron compounds for the X-ray radiation and the lower zeolite content in the samples by the addition of iron salts. [25] The stoichiometric amount of Fe for ion exchange to form

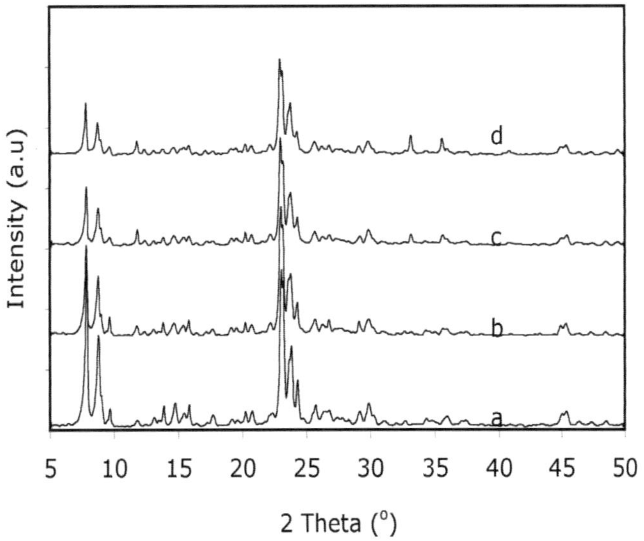

Figure 3. XRD profiles of Fe/ZSM-5 catalysts prepared with $FeCl_2$ precursor. a) 1%Fe/ZSM-5; b) 2.5%Fe/ZSM-5; c) 5%Fe/ZSM-5; d) 7.5%Fe/ZSM-5.

Fe-ZSM-5 was 5%. Interestingly, when the loading of Fe was less than 5%, no peak for α-Fe_2O_3 specie was observed, whereas when the Fe loading was over 5%, two additional lines were detected at 33.2 and 35.5°. These lines were due to the strongest lines of α-Fe_2O_3. From the XRD it can be seen that although the iron source was $FeCl_2$, after calcination, the excess iron species became iron oxides. This result is also an indication that a solid exchange process has occurred (as well as the hydrolysis of $FeCl_2$) during the the preparation/calcination procedure, so α-Fe_2O_3 was detected when the iron loading was beyond the stoichiometric amount of Fe for solid exchange. Based on the catalytic activity results, the activity increased with increasing Fe loading, but the change was not much when the Fe loading was >5%. This result indicates that the crystalline α-Fe_2O_3 contributed little to the SCR activity. No $FeCl_2$ was detected in all the prepared catalysts. From the BET surface areas and pore volumes of catalysts (not shown) with different iron loadings, it was found that both BET surface area and pore volume decreased with increasing iron loading, which is attributed to the blockage of the precipitated Fe species formed by the excess iron. It could also be concluded that the ion exchange was incomplete because even in the 5%Fe/ZSM-5 sample α-Fe_2O_3 was detected by XRD. Further evidences for ion exchange taking place during the incipient wetness impregnation and calcination follow.

Since ESR is a powerful technique for analyzing the state of ferric ions, [29-33] the catalysts were characterized by ESR. Figure 4 shows the ESR spectra taken at 20°C of Fe-ZSM-5 prepared by aqueous ion exchange, fresh and used 7.5%Fe/ZSM-5 samples prepared by impregnation methods. The "used" sample was that after subjecting to SCR reaction at 400°C. A strong line at g = 4.3 was observed on Fe-ZSM-5 prepared by aqueous ion exchange. This signal can be assigned to Fe^{3+} in tetrahedral coordination. [29-33] Another weak broad line was also detected at g = 2.2 (not shown) which probably came from aggregated Fe^{3+} ions in the extraframework α-Fe_2O_3 species exhibiting ferromagnetic behavior. [31] The fresh Fe/ZSM-5 sample prepared by impregnation shows

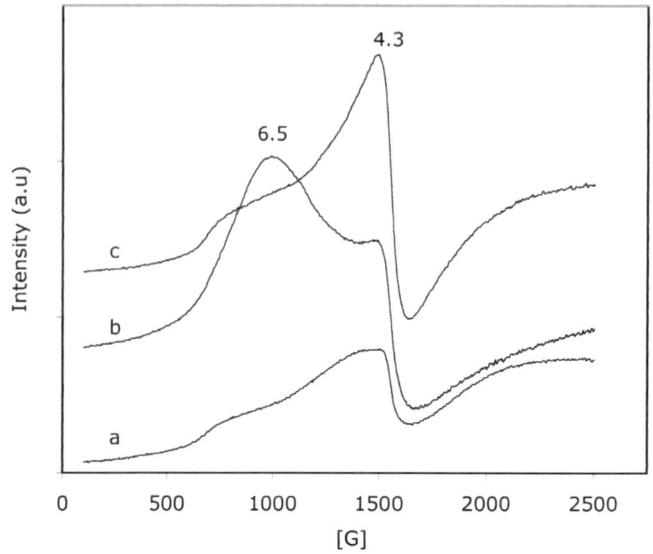

Figure 4. EPR spectra of Fe/ZSM-5 catalysts at room temperature. a) Fe-ZSM-5 prepared by aqueous ion-exchange; b) Fresh 7.5%Fe/ZSM-5; c) 7.5%Fe/ZSM-5 (used) after SCR reaction at 400 °C.

a relatively small signal at g = 4.3 and a strong signal at g = 6.5, the latter signal being contributed to the distorted tetrahedral coordination. [29-33] A substantial further rise of the broad line at g = 2.2 also took place. For the used sample, a significantly change of lines at g = 4.3 and g = 6.5 occurred. The intensity of the line at g = 4.3 increased significantly while the intensity of the line at g = 6.5 nearly disappeared. It is not surprised because the interaction of coordinatively unsaturated Fe^{3+} ions with such strong ligands as ammonia or pyridine resulted in an increase in field symmetry, provoking the shift of ESR lines with g = 6.25 to a line with g = 4.27. [30, 32] The g = 2.2 signal remained unchanged for this used sample. Figure 5 shows the XPS spectra of $Fe2p_{3/2}$ on the fresh and used 7.5%Fe/ZSM-5 catalysts. Two distinct bands centered at 711.7 and 708 eV were observed on the fresh sample. These values are close to the range of the $Fe2p_{3/2}$ binding energy of iron in Fe_2O_3 and FeO, [34] indicating that the iron in the Fe/ZSM-5 were present in two valences of +3 and +2, respectively. On the fresh Fe-ZSM-5 prepared by aqueous ion exchange, only one XPS band centered at 711.7 eV was observed. [35] On the used 7.5%Fe/ZSM-5 sample, two XPS bands were observed at the same positions to the fresh sample, but the intensity of the band centered at 708 eV decreased significantly, indicating Fe^{2+} was partially oxidized to Fe^{3+} during the SCR reaction. In the ESR measurement, Fe^{2+} species was not detected because the Fe ions observed in various zeolite systems are usually in the high spin $3d^5$ configuration in ESR.

We have shown that the proposed impregnation method based on the NH_4-ZSM-5 and $FeCl_2$ is a surprisingly simple and effective process for the preparation of Fe/ZSM-5 catalyst, which has superior catalytic activities for the SCR of NO with ammonia in the presence of excess oxygen compared to all other known catalysts. This process can be very useful for large-scale production of the catalyst. This process is in effect a combination of the conventional impregnation and solid-state exchange processes but the gas atmosphere for calcination is not inert gas. Another process in which the Fe/ZSM-5 sample was first calcined in an inert gas (He) instead of air and then calcined in air was also investigated, and the activities of the catalysts prepared by these two processes were

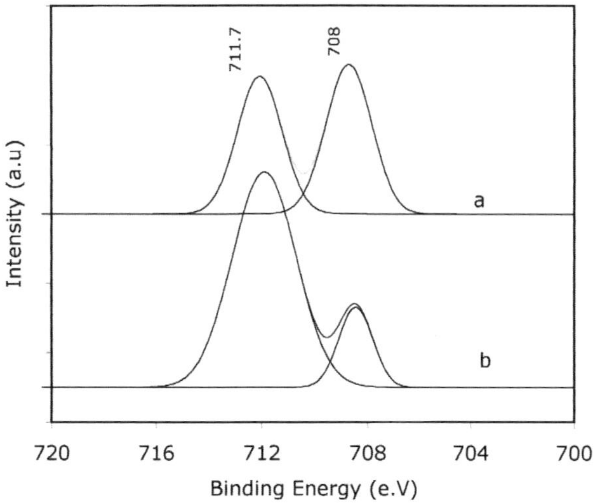

Figure 5. XPS spectrum of $Fe2p_{3/2}$ on 7.5%Fe/ZSM-5 catalysts. a) Fresh catalyst; b) After SCR reaction at 400 °C for 5 hrs;

very similar, which indicated that even in the air the exchange between NH_4^+ and Fe^{2+} took place as evidenced by the ESR results. From the XRD and ESR, however, it could be seen that the exchange was incomplete. Such incomplete exchange should lead to produce the Brønsted type acid in the catalysts after the decomposition of NH_4-ZSM-5, which is considered as the active site for ammonia adsorption. Based on the catalytic activity and characterization results, it can be concluded that the Fe^{2+} species may contribute to the high activities especially in the lower temperature range.

References

[1] H. Bosch, and F. Janssen, *Catal. Today* **1988**, 2, 369.
[2] G. Busca, L. Lietti, G. Ramis, and F. Berti, *Appl. Catal. B* **1998**, 18, 1.
[3] J.A. Odriozola, H. Heinemann, G.A. Somorjai, J.E. Garcia de la Banda, P. Pereira, *J. Catal.* **1989**, 119, 71.
[4] G. Ramis, G. Busca, F. Bregani, P. Forzatti, *Appl. Catal.* **1990**, 64, 259.
[5] M. Schramlmarth, A. Wokaun, A. Baiker, *J. Catal.* **1990**, 124, 86.
[6] G. Went, L.-J. Leu, R.R. Rosin, A.T. Bell, *J. Catal.* **1992**, 134, 492.
[7] U.S. Ozkan, Y. Cai, M.W. Kumthekar, *J. Catal.* **1994**, 149, 390.
[8] N.-Y. Topsøe, J.A. Dumesic, H. Topsøe, *J. Catal.* **1995**, 151, 241.
[9] J.R. Kiovsky, P.B. Koradia, and C.T. Lim, *I&EC Prod. Res. Dev.* **1980**. 19, 218.
[10] S. Sato, H. Hirabayashi, H. Yahiro, N. Mizuno, and M. Iwamoto, *Catal. Lett.* **1992**, 12, 193.
[11] X. Feng, and W.K. Hall, *J. Catal.* **1997**, 166, 368.
[12] W.K. Hall, X. Feng, J. Dumesic, and R. Watwe, *Catal. Lett.* **1998**, 52, 13.
[13] H.-Y. Chen, and W.M.H. Sachtler, *Catal. Today* **1998**, 42, 73.
[14] R.Q. Long, and R.T. Yang, *J. Am. Chem. Soc.* **1999**, 121, 5595.
[15] R.Q. Long, and R.T. Yang, *J. Catal.* **1999**, 188, 332.
[16] A.Z. Ma, and W. Grunert, *Chem. Commun.* **1999**, 71.
[17] J. Valyon, W.S. Millman, and W.K. Hall, *Catal. Lett.* **1994**, 24, 215.

[18] Q. Sun, Z.-H. Gao, H.-Y. Chen, W.M.H. Sachtler, *J. Catal.* **2001**, 201, 89.
[19] M.D. Amiridis, F. Puglisi, J.A. Dumesic, W.S. Millman, N.-Y. Topsøe, *J. Catal.* **1993**, 142, 572.
[20] M. Kögel, V.H. Sandoval, W. Schwieger, A. Tissler, T. Turek, *Catal. Lett.* **1998**, 51, 23.
[21] R. Joyner, M. Stockenhuber, *J. Phys. Chem. B* **1999**, 103, 5963.
[22] L.-J. Lobree, I.-C. Hwang, J.A. Reimer, A.T. Bell, *J. Catal.* **1999**, 186, 242.
[23] E.-M. El-Malki, R.A. van Santen, W.M.H. Sachtler, *J. Catal.* **2000**, 196, 212.
[24] K. Hadjiivanov, H. Knözinger, B. Tsyntsarski, and L. Dimitrov, *Catal. Lett.* **1999**, 62, 35.
[25] M. Rauscher, K. Kesore, R. Mönnig, W. Schwieger, A. Tißler, T. Turek, *Appl. Catal. A* **1999**, 184, 249.
[26] R.Q. Long, and R.T. Yang, *Catal. Lett.* **2001**, 74, 201.
[27] R. Chang and S. Cardoso, *Advanced NOx Catalyst Development Update*, EPRI, Palo Alto, CA, and, Oglethorpe Power Corporation, Juliette, GA, EPRI Report 1006657 (2003).
[28] L.S. Cheng, R.T. Yang, and N. Chen, *J. Catal.* **1996**, 164, 70.
[29] A.V. Kucherov and M. Shelef, *J. Catal.* **2000**, 195, 106.
[30] A.A. Slinkin, and A.V. Kucherov, *Catal. Today* **1997**, 36, 485.
[31] A.V. Kucherov, C.N. Montreuil, T.N. Kucherova, and M. Shelef, *Catal. Lett.* **1998**, 56, 173.
[32] A.V. Kucherov and A.A. Slinkin, *Zeolites* **1988**, 8, 10.
[33] D. Goldfarb, M. Bernardo, K.G. Strohmaier, D.E.W. Vaughan and H. Thomann, *J. Am. Chem. Soc.* **1994**, 116, 6344.
[34] C.D. Wagner, W.M. Riggs, L.E. Davis, J.F. Moulder, and G.E. Mullerberg, Eds., " *Handbook of X-Ray Photoelectron Spectroscopy.*" Perkin-Elmer Corp. Eden Prairie, **1979**.
[35] R.Q. Long, and R.T. Yang, *J. Catal.* **2000**, 194, 80.

Paper Number: _________

A Superior Fe/ZSM-5 Catalyst for Selective Catalytic Reduction (SCR) of Nitric Oxide with Ammonia

By

Gongshin Qi and Ralph T. Yang

Paper Presented at the Spring National AICHE Meeting, April 25-29, 2004

Session: Application of Process Integration Tools for Energy Efficient Design and Water Minimization Design

UNPUBLISHED

Current Affiliation and address:
Department of Chemical Engineering, University of Michigan
Ann Arbor, MI 48109

Abstract

- We have shown that the proposed impregnation method based on the NH_4-ZSM-5 and $FeCl_2$ is a surprisingly simple and effective process for the preparation of Fe/ZSM-5 catalyst, which has superior catalytic activities for the SCR of NO with ammonia in the presence of excess oxygen compared to all other known catalysts. This process can be very useful for large-scale production of the catalyst. This process is in effect a combination of the conventional impregnation and solid-state ion-exchange processes. Our previous results showed that although iron cations were exchanged with NH_4-ZSM-5 as Fe^{2+}, iron cations in the Fe/ZSM-5 were present as both Fe^{3+} and Fe^{2+} after calcination while for Fe-ZSM-5 catalyst, most iron cations were present as Fe^{3+}. The co-existence of Fe^{3+} and Fe^{2+} metal cations in the catalyst may be beneficial to oxygen adsorption and activation. Therefore, a high activity for SCR is expected. the Fe/ZSM-5 should be a promising catalyst for the SCR process. In addition to the activity test, this catalyst was also characterized by XRD, EPR, TPR, FTIR and TEM.

A Superior Fe/ZSM-5 Catalyst for Selective Catalytic Reduction (SCR) of Nitric Oxide with Ammonia

Gongshin Qi
and
Dr. Ralph T. Yang

University of Michigan
Department of Chemical Engineering
Ann Arbor, MI 48109

Agenda

- Introduction
- Experimental
  - ✓Catalysts Preparation, Characterization
  - ✓Activity Test
- Results and Discussion
- Summary
- Questions

Introduction

- ✓ Nitrogen oxides (NO, NO_2 and N_2O) are major source for air pollution.
- ✓ Photochemical smog, acid rain, ozone depletion, and greenhouse effects.
- ✓ The current technology for reducing nitrogen oxides emissions from power plants is selective catalytic reduction (SCR) of NO_x (x=1, 2) with ammonia in excess oxygen.
- ✓ H-zeolite and ion-exchanged molecular sieves have received much attention for SCR of NO_x by both hydrocarbon and ammonia in recent years.
- ✓ Fe-ZSM-5 showed the highest activities among all known catalysts for the reaction.
- ✓ In this work, we developed a new and very simple method for preparing Fe/ZSM-5 catalyst by conventional incipient-wetness impregnation method using the NH_4-ZSM-5 and $FeCl_2$ as starting materials. The resulting catalyst prepared from $FeCl_2$ yielded significantly higher activities than catalysts prepared by all other methods.

Experimental

- ✓ An aqueous solution of $FeCl_2$ having the same volume as the total pore volume of the zeolite was prepared.
- ✓ Solution was spontaneously imbibed by the porous zeolite. The sample was then calcined in air at 500°C for 6 hours.
- ✓ Catalyst was characterized by XRD, ESR and XPS.
- ✓ SCR Experiment: The NO and NO_2 concentrations were continually monitored by a chemiluminescent NO/NO_x analyzer. The products were also analyzed by a gas chromatograph (Shimadzu, 8A) at 50 °C with 5A molecular sieve column for N_2 and Porapak Q column for N_2O.

Experimental

Preparation routes

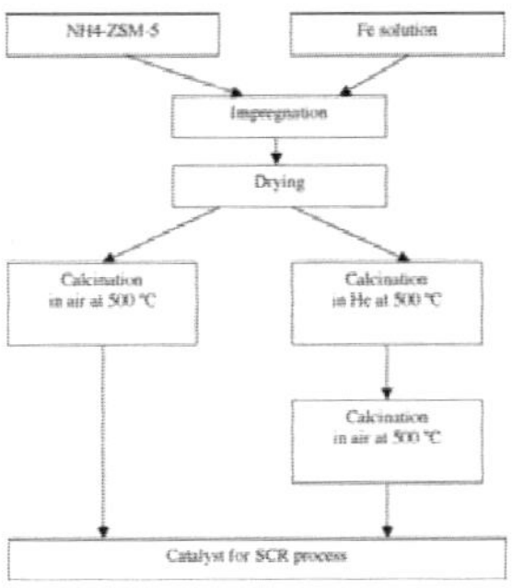

Experimental

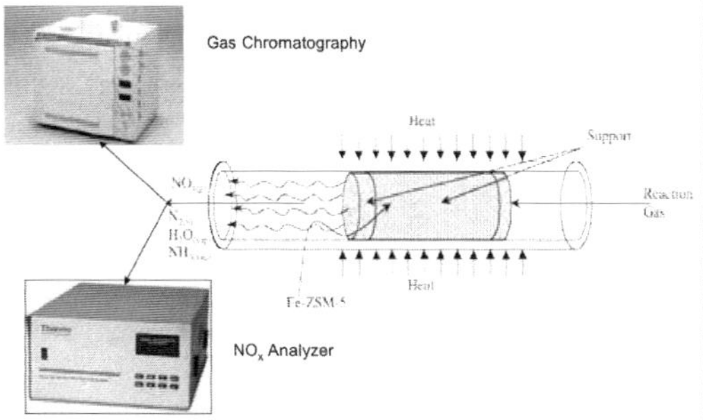

Fe-ZSM-5 XRD

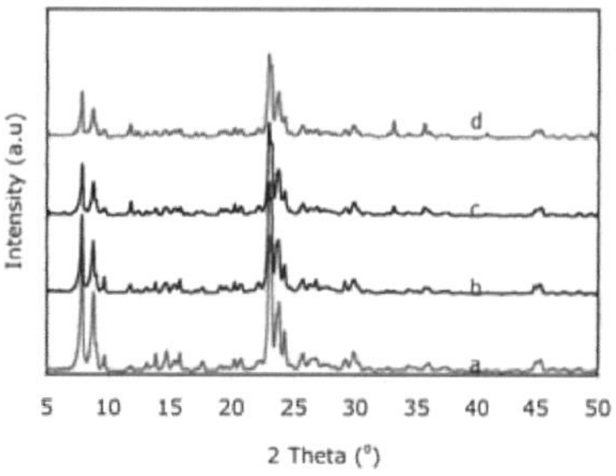

XRD profiles of Fe/ZSM-5 catalysts prepared with $FeCl_2$ precursor.
a) 1%Fe/ZSM-5; b) 2.5%Fe/ZSM-5; c) 5%Fe/ZSM-5; d) 7.5%Fe/ZSM-5.

Fe-ZSM-5 XRD

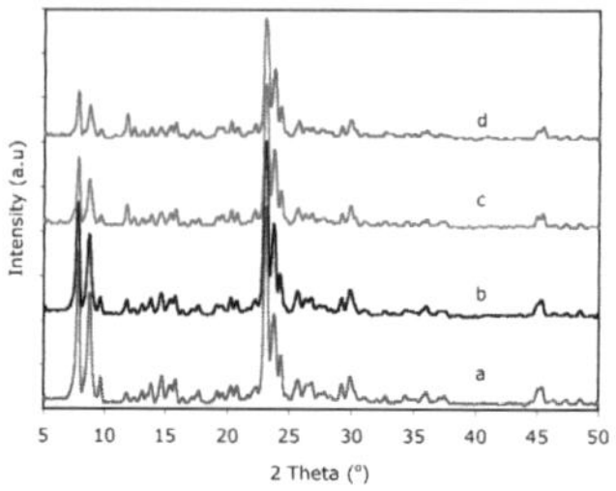

XRD profiles of Fe/ZSM-5 catalysts prepared by two-step calcinations with $FeCl_2$ as the precursor. a) 1%Fe/ZSM-5; b) 2.5%Fe/ZSM-5; c) 5%Fe/ZSM-5; d) 7.5%Fe/ZSM-5.

Fe-ZSM-5 TPR

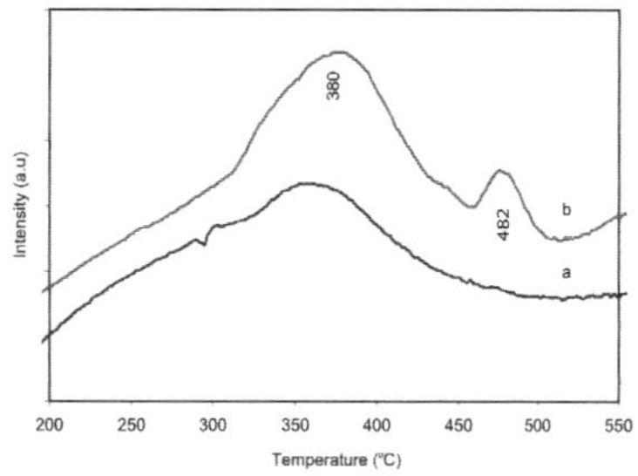

H_2-TPR profiles of Fe/ZSM-5 catalysts prepared by two-step method.
a) 5%Fe/ZSM-5, b) 7.5%Fe/ZSM-5.

Fe-ZSM-5 XPS

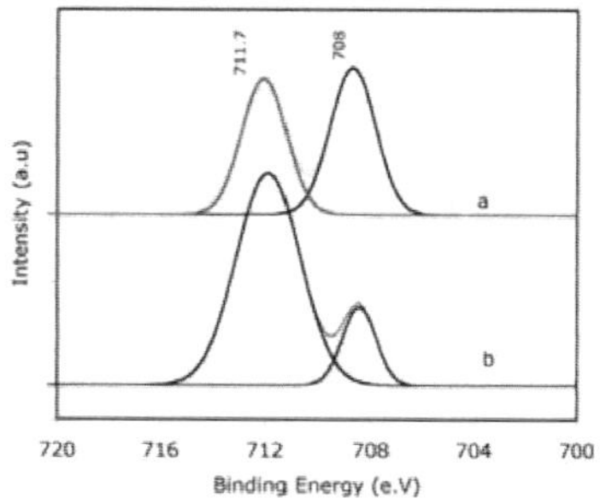

XPS spectrum of $Fe2p_{3/2}$ on 7.5%Fe/ZSM-5 catalysts
a) Fresh catalyst; b) After SCR reaction at 400 °C for 5 hrs.

Fe-ZSM-5 EPR

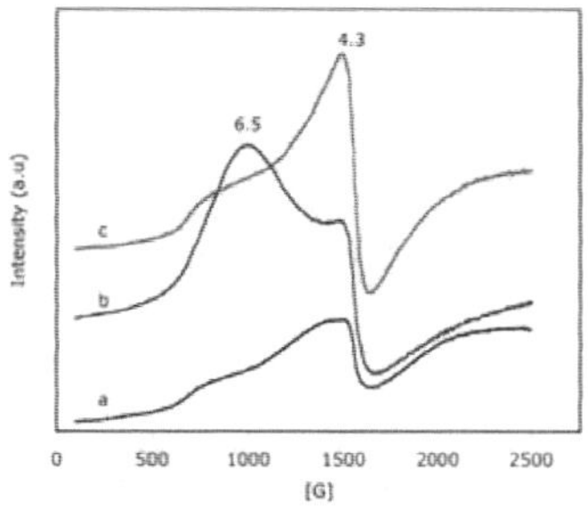

EPR spectra of Fe/ZSM-5 catalysts at room temperature
a) Fe-ZSM-5 prepared by aqueous ion-exchange; b) Fresh 7.5%Fe/ZSM-5;
c) 7.5%Fe/ZSM-5 (used) after SCR reaction at 400 °C.

Fe-ZSM-5 TEM

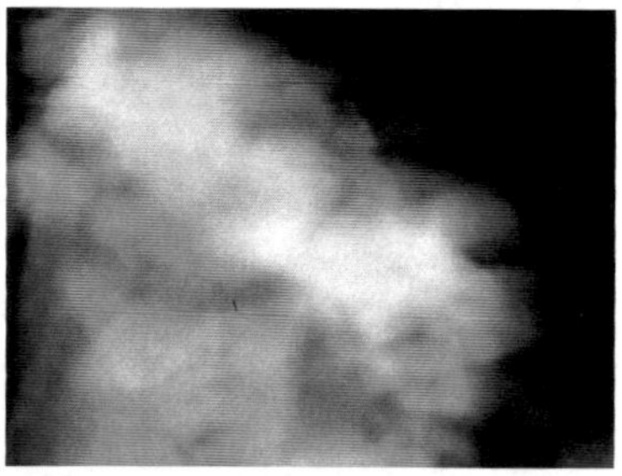

EPR spectra of Fe/ZSM-5 catalysts at room temperature
a) Fe-ZSM-5 prepared by aqueous ion-exchange; b) Fresh 7.5%Fe/ZSM-5;
c) 7.5%Fe/ZSM-5 (used) after SCR reaction at 400 °C.

Fe-ZSM-5 TEM

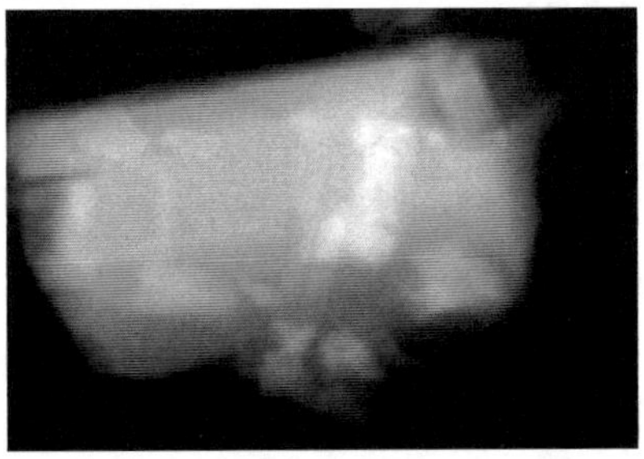

EPR spectra of Fe/ZSM-5 catalysts at room temperature
a) Fe-ZSM-5 prepared by aqueous ion-exchange; b) Fresh 7.5%Fe/ZSM-5;
c) 7.5%Fe/ZSM-5 (used) after SCR reaction at 400 °C.

Fe-ZSM-5 Activity

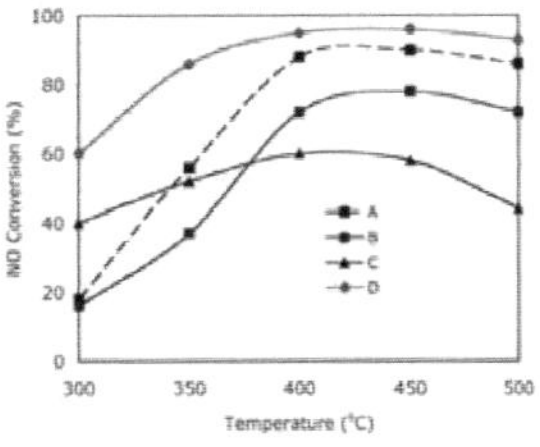

Catalytic activities for NO reduction by ammonia on different catalysts.
Reaction conditions: 40 mg catalyst, 1000 ppm NO, 1000 ppm NH_3, 2% O_2, balance He,
and GHSV = 5.7 ∅10⁵ h⁻¹(ambient conditions)
A) Fe-ZSM-5, prepared by ion-exchange in $FeCl_2$ solution;
B) 2.5%Fe/ZSM-5, prepared by impregnation with $Fe(NO_3)_3$ as precursor;
C) 4.4%V_2O_5-8.2%WO_3/TiO_2 prepared by impregnation;
D) 2.5%Fe/ZSM-5, prepared by impregnation with $FeCl_2$ as precursor.

Fe-ZSM-5 Activity (Cont.)

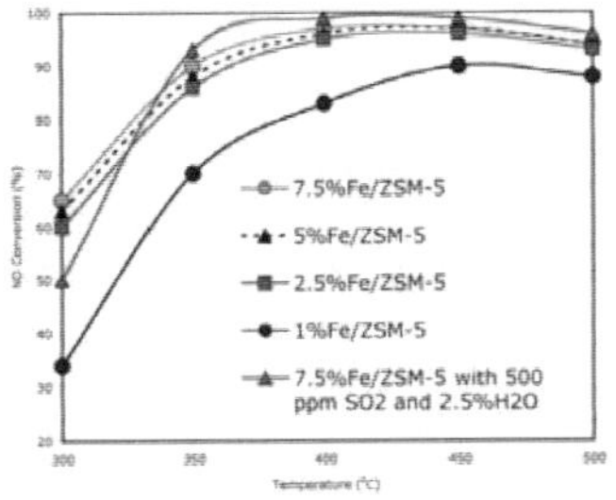

Catalytic activities for NO reduction by ammonia on Fe/ZSM-5 catalysts with different Fe loadings.
Reaction conditions: 40 mg catalyst, 1000 ppm NO, 1000 ppm NH_3, 2% O_2, 500 ppm SO_2 (when used), 2.5% H_2O (when used), balance He, and GHSV = 5.7 ∅10⁵ h⁻¹.

Fe-ZSM-5 Activity (Cont.)

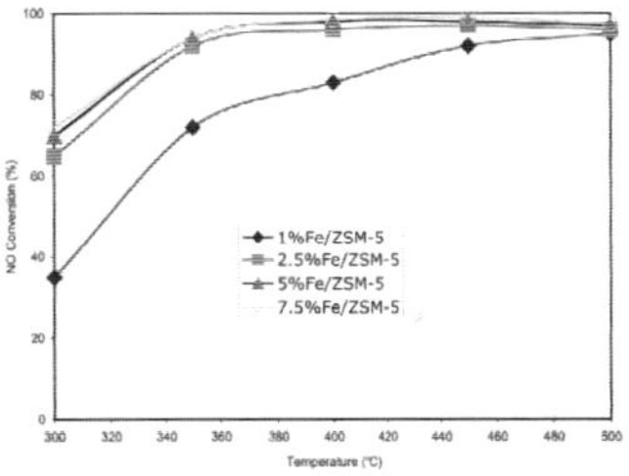

Catalytic activities for NO reduction by ammonia on Fe/ZSM-5 catalysts prepared by 2 step method with different Fe loadings.
Reaction conditions: 40 mg catalyst, 1000 ppm NO, 1000 ppm NH_3, 2% O_2, 500 ppm SO_2 (when used), 2.5% H_2O (when used), balance He, and GHSV = 5.7 ∅10⁵ h⁻¹.

Fe-ZSM-5 FTIR

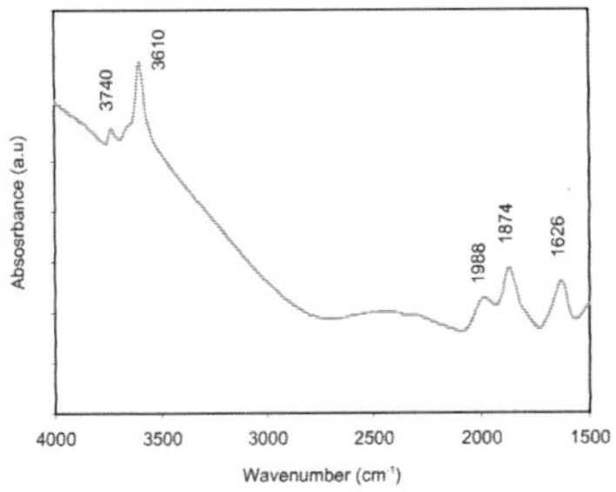

FTIR spectrum of 2.5%Fe/ZSM-5 catalyst at 30 °C with empty cell as background.

Fe-ZSM-5 FTIR

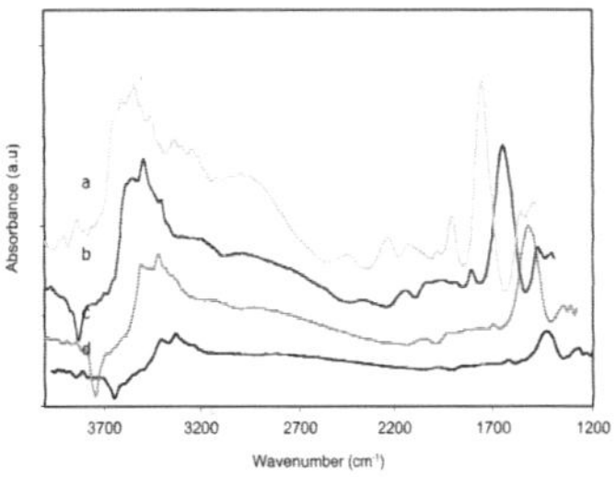

FTIR spectra taken upon passing in He flow over the presorbed NH_3 Fe/ZSM-5 for a) 30, b) 100, c) 200, d) 300 °C.

Fe-ZSM-5 FTIR

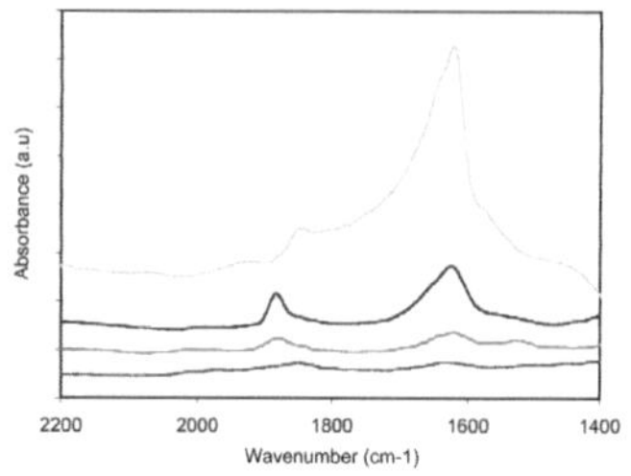

FTIR spectra taken upon passing in NO $+O_2$ flow over the Fe/ZSM-5 for a) 30, b) 100, c) 200, d) 300 °C.

Fe-ZSM-5 FTIR

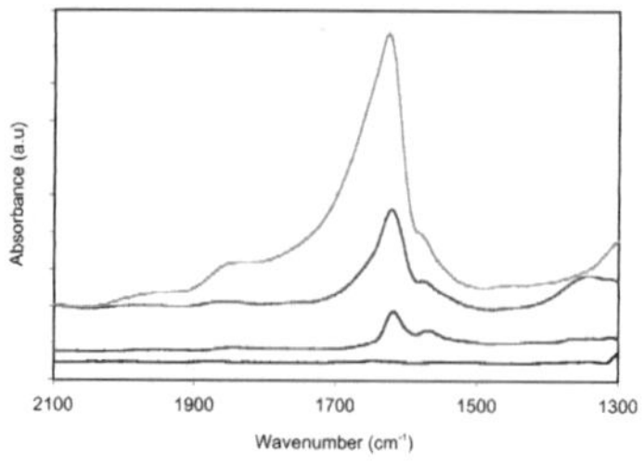

FTIR spectra taken upon passing in He flow over the NO + O_2 presorbed Fe/ZSM-5 for a) 30, b) 100, c) 200, d) 300 °C.

Fe-ZSM-5 FTIR

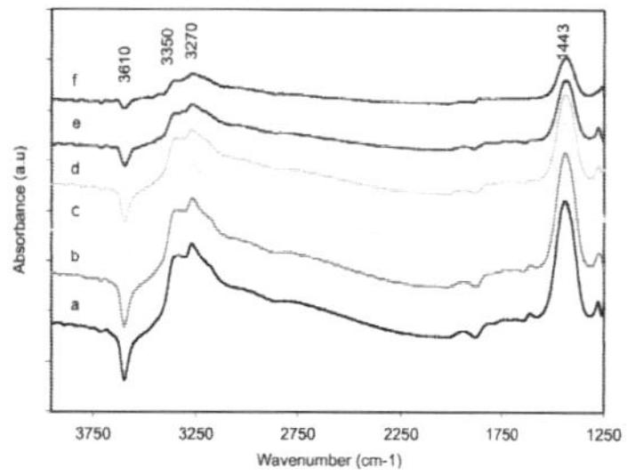

FTIR spectra taken at 200 °C upon passing 1000 ppm NO + 2% O_2 over the NH_3 presorbed on Fe/ZSM-5 for a) 0, b) 0.5, c) 1.5, d) 3, e) 5, f) 8 min.

Summary

- ✓ The proposed impregnation method based on the NH_4-ZSM-5 and $FeCl_2$ is a surprisingly simple and effective process for the preparation of Fe/ZSM-5 catalyst, which has superior catalytic activities for the SCR of NO with ammonia in the presence of excess oxygen compared to all other known catalysts.
- ✓ Another process in which the Fe/ZSM-5 sample was first calcined in an inert gas (He) instead of air and then calcined in air was also investigated, and the activities of the catalysts prepared by these two processes were very similar, which indicated that even in the air the exchange between NH_4^+ and Fe^{2+} took place as evidenced by the ESR results. From the XRD and ESR, however, it could be seen that the exchange was incomplete. Such incomplete exchange should lead to produce the Br∅nsted type acid in the catalysts after the decomposition of NH_4-ZSM-5, which is considered as the active site for ammonia adsorption. Based on the catalytic activity and characterization results, it can be concluded that the Fe^{2+} species and the highly dispersed iron oxides may contribute to the high activities especially in the lower temperature range.